Lecture Notes of the Institute for Computer Sciences, Social Informatics and Telecommunications Engineering 174

More information about this series at http://www.springer.com/series/8197

Julian Cheng · Ekram Hossain
Haijun Zhang · Walid Saad
Mainak Chatterjee (Eds.)

Game Theory
for Networks

6th International Conference, GameNets 2016
Kelowna, BC, Canada, May 11–12, 2016
Revised Selected Papers

 Springer

Editors
Julian Cheng
School of Engineering
University of British Columbia
Kelowna, BC
Canada

Ekram Hossain
University of Manitoba
Winnipeg, MB
Canada

Haijun Zhang
University of British Columbia
Vancouver, BC
Canada

Walid Saad
Wireless@VT, Bradley Department of ECE
Virginia Tech
Blacksburg, VA
USA

Mainak Chatterjee
Department of Electrical Engineering
and Computer Science
University of Central Florida
Orlando, FL
USA

ISSN 1867-8211   ISSN 1867-822X (electronic)
Lecture Notes of the Institute for Computer Sciences, Social Informatics
and Telecommunications Engineering
ISBN 978-3-319-47508-0   ISBN 978-3-319-47509-7 (eBook)
DOI 10.1007/978-3-319-47509-7

Library of Congress Control Number: 2016957479

Printed on acid-free paper

This Springer imprint is published by Springer Nature
The registered company is Springer International Publishing AG
The registered company address is: Gewerbestrasse 11, 6330 Cham, Switzerland

Preface

The 6th International Conference on Game Theory for Networks (Gamenets) was held during May 11–12, 2016, on the UBC-Okanagan campus in spectacular Kelowna, Canada. Kelowna is widely recognized as one of the world's most liveable cities. The mission of the conference is to share novel basic research ideas as well as experimental applications in the Gamenets area in addition to identifying new directions for future research and development.

Gamenets 2016 had 13 peer-reviewed papers and a plenary talk on "Social Learning and Social Sensing" by Dr. Vikram Krishnamurthy of the University of British Columbia.

We would like to thank the authors for providing the content of the program. We would also like to express our gratitude to the Technical Program Committee (TPC) and reviewers, who worked very hard on reviewing the papers. This year, we received 26 paper submissions from authors all over the world. After a rigorous peer review by the TPC, 13 papers were accepted.

We would like to thank our financial sponsor EAI (European Alliance for Innovation) for their support in making Gamenets 2016 a successful event.

October 2016   Julian Cheng

Organization

Organizing Committee

Honorary Chairs

Victor C.M. Leung	University of British Columbia, Canada
Imrich Chlamtac	Creat-Net, Italy

Conference General Chairs

Julian Cheng	University of British Columbia, Canada
Ekram Hossain	University of Manitoba, Canada
Haijun Zhang	University of British Columbia, Canada

Technical Program Committee Co-chairs

Walid Saad	Virginia Tech, USA
Mainak Chatterjee	University of Central Florida, USA

Workshops Committee Chair

Zhengguo Sheng	University of Sussex, UK

Publications Committee Chair

Weisi Guo	University of Warwick, UK

Publicity and Social Media Chair

Chunxiao Jiang	Tsinghua University, China

Sponsorship and Exhibits Committee Chair

Hui Ma	University of British Columbia, Canada

Local Arrangements Committee Chairs

Md. Zoheb Hassan	University of British Columbia, Canada
Fang Fang	University of British Columbia, Canada

Website Committee Chair

Yanjie Dong	University of British Columbia, Canada

Steering Committee

Imrich Chlamtac	Create-Net, Italy
Athanasios Vasilakos	Kuwait University, Kuwait

Technical Program Committee

Mingyan Liu	University of Michigan, USA
Charles Kamhoua	Air Force Research Laboratory, USA
Ning Wang	Zhengzhou University, China
Carlos Cid	University of London, UK
Mohammad Hossein Manshaei	Isfahan University of Technology, Iran
Emmanouil Manos Panaousis	University of Brighton, UK
Dapeng Li	Nanjing University of Posts and Telecommunications, China
Sumudu Samarakoon	University of Oulu, Finland
Shamik Sengupta	University of Nevada, USA
Luca Sanguinetti	University of Pisa, Italy
Kenza Hamidouche	CentraleSupelec, France
Anibal Sanjab	Virginia Tech, USA
Lingjie Duan	Singapore University of Technology and Design, Singapore
Yuan Luo	Chinese University of Hong Kong, SAR China
Wen Ji	The Institute of Computing Technology, Chinese Academy of Science, China
Lin Gao	Chinese University of Hong Kong, SAR China
Yuhua Xu	PLA University of Science and Technology, China
Chungang Yang	Xidian University, China
Husheng Li	The University of Tennessee, USA
Daqiang Zhang	Tongji University, China

Contents

Design and Analysis of Economic Games

Algorithmic Game Theory

Strategic Seeding of Rival Opinions

Samuel D. Johnson[1]([✉]), Jemin George[2], and Raissa M. D'Souza[3]

[1] HRL Laboratories, LLC, Malibu, CA 90265, USA
sdjohnson@hrl.com
[2] United States Army Research Laboratory, Adelphi, MD 20783, USA
jemin.george.civ@mail.mil
[3] University of California, Davis, CA 95616, USA
raissa@cse.ucdavis.edu

Abstract. We present a network influence game that models players strategically seeding the opinions of nodes embedded in a social network. A social learning dynamic, whereby nodes repeatedly update their opinions to resemble those of their neighbors, spreads the seeded opinions through the network. After a fixed period of time, the dynamic halts and each player's utility is determined by the relative strength of the opinions held by each node in the network *vis-à-vis* the other players. We show that the existence of a pure Nash equilibrium cannot be guaranteed in general. However, if the dynamics are allowed to progress for a sufficient amount of time so that a consensus among all of the nodes is obtained, then the existence of a pure Nash equilibrium can be guaranteed. The computational complexity of finding a pure strategy best response is shown to be NP-complete, but can be efficiently approximated to within a $(1 - 1/e)$ factor of optimal by a simple greedy algorithm.

Keywords: Social networks · Opinion dynamics · Game theory · Nash equilibrium · Computational complexity · Approximation algorithm

1 Introduction

Opinions are shaped by the information individuals obtain through their social connections. These opinions inform our career decisions, political views, and purchasing behaviors, which can ultimately spread to affect an entire society. The *influence maximization problem* [14] asks: Given the ability to seed a small number of individuals (nodes) in a social network to adopt a desired behavior (*e.g.*, to purchase a product, support a political candidate, contract an infection, etc.), which nodes should be selected so that the behavior subsequently spreads to a maximum fraction of the entire population?

Strategic extensions to the influence maximization problem that involve two or more players representing competing, substitutable behaviors (products, opinions, infections, etc.) where each player selects a set of seed nodes so as to maximize the fraction of the population that eventually adopts their represented

This research was conducted while S.D. Johnson was a graduate student at the University of California, Davis.

© ICST Institute for Computer Sciences, Social Informatics and Telecommunications Engineering 2017
J. Cheng et al. (Eds.): GameNets 2016, LNICST 174, pp. 3–12, 2017.
DOI: 10.1007/978-3-319-47509-7_1

behavior, have subsequently been studied; see, for example, [1,4,5,8,11,17]. A common feature among most of these models is that the diffusion dynamics they use involve nodes making binary decisions to determine whether to fully adopt one of the diffused opinions. Furthermore, these models often employ progressive SI-style dynamics in which nodes that do not have an opinion must first make an irreversible decision to fully commit themselves to a single opinion in order to become a participant in the subsequent propagation of their adopted opinion. Although such dynamics meet the desiderata of many applications with practical importance, we believe them to be unnatural for the diffusion of opinions through social networks because of the requirement that opinions are only allowed to pass through nodes that have already committed irreversibly to a particular preference.

In the current work, we present the *network influence game* (NIG) to model the strategic seeding of opinions in a social network by $m \geq 2$ players and employs a dynamic that is arguably more appropriate than the SI-style dynamics used elsewhere. In our model, each node maintains a vector of opinions toward the m players, and in each discrete time step, a node's opinion is updated to reflect a weighted averaging of their previous opinion and the opinions of their neighbors. These dynamics proceed in accordance with the *consensus dynamic* [6], and terminate after a period of T time steps. The consensus dynamic has played a prominent role in the study of opinion dynamics in social networks – see, for example, the surveys by Jackson [12,13]. In our competitive model, players' utilities are defined to be a function of each node's relative opinions toward the players upon the conclusion of the dynamic process.

The NIG model is formally presented in Sect. 2, followed by a discussion on the nature of influence pertaining to it in Sect. 2.1. Section 3 contains our results on the existence of pure Nash equilibrium strategies. We show that existence can be guaranteed if the dynamics run until a consensus opinion is reached, but cannot be guaranteed otherwise. In Sect. 4 we show that computing a pure strategy best response is NP-complete, yet can be efficiently approximated to within a factor of $(1 - 1/e)$ of optimal. Section 5 concludes the paper with a discussion and suggested topics for future research.

2   Model

The *network influence game* NIG is specified by a tuple $\langle M, G, \Delta_C, T, \mathbf{b}, \pi \rangle$ with a player set, $M = \{1, \ldots, m\}$; a network, $G = (V, E)$, represented by a weighted digraph with $|V| = n$ nodes; the opinion dynamic, Δ_C, that determines how influence spreads between adjacent nodes in the network; the length of time, $T \in \mathbb{N}$, that the diffusion dynamic is allowed to proceed; a profile, $\mathbf{b} = (b_1, \ldots, b_m)$, of integer seed budgets $b_i > 0$ for each player $i \in M$; and a utility function, $\pi_i(\cdot)$, that aggregates the nodes' opinions upon the conclusion of the diffusion dynamics at time T into a non-negative payoff for each player $i \in M$.

We assume that the graph G is strongly connected. Furthermore, we stipulate that for all edges $(u, v) \in E$, the edge weight $w(u, v) > 0$ and, for all nodes $v \in V$, the sum of all incoming edge weights equals one.

A (pure) strategy for player i is a subset $s_i \subseteq V$ of at most b_i seed nodes. A strategy profile $s = (s_1, \ldots, s_m)$ specifies the seed nodes chosen by all m players, and is used to set the initial conditions for the opinion dynamics, the result of which determines the players' utilities.

The NIG's opinion dynamics, Δ_C, is based on DeGroot's *consensus dynamic* [6]. For this dynamic, each node $v \in V$ maintains a length-m opinion vector $\mathbf{x}^v = (x_1^v, \ldots, x_m^v)$ with the entry $0 \leq x_i^v \leq 1$ representing v's opinion toward player i. If $x_i^v > x_j^v$, then this is to be interpreted as v holding a more favorable opinion toward player i than toward player j. We require that the sum of a node's opinions is at most one; $|\mathbf{x}^v|_1 = \sum_{i \in M} x_i^v \leq 1$. Since we need to refer to the evolution of a node's opinion over time, we will use $\mathbf{x}^v(t)$ to denote v's vector of opinions at time t, with entry $x_i^v(t)$ representing v's opinion toward player i at time t.

The diffusion process is initialized by s at time $t = 0$. This involves each player i implanting a "seed opinion" $\mathbf{y}^i$ into each of the nodes included in s_i. The seed opinions $\mathbf{y}^i = (y_1^i, \ldots, y_m^i)$ are defined as $y_i^i = 1$ and $y_j^i = 0$ for all $j \neq i$, meaning that $\mathbf{y}^i$ specifies a high (maximum) opinion toward player i and a low (minimum) opinion toward all other players j. Let $M_v(s) = \{i \mid v \in s_i\} \subseteq M$ denote the subset players that include a given node v in their strategy, and define $m_v(s) = |M_v(s)|$. Finally, let V_s denote the subset of nodes that are designated as seed nodes by at least one player. We initialize the opinions of the nodes $v \in V$ towards each player $i \in M$

$$x_i^v(0) = \begin{cases} \frac{1}{m_v(s)} \sum_{j \in M_v} y_i^j & \text{if } v \in V_s \\ \varepsilon & \text{if } v \in V \setminus V_s, \end{cases} \qquad (1)$$

where $0 < \varepsilon \ll 1/m$ is a small constant. Equation (1) specifies that each seed node $v \in V_s$ is initialized to the average seed opinion of the players that include v in their strategies. Otherwise, for a node $v \notin V_s$, initialization involves assigning a ε opinion value toward every player $i \in M$.

The consensus dynamic, Δ_C, proceeds in discrete time steps $t = 1, 2, \ldots, T$ and specifies that the opinion of a node v at time t is a weighted average of its prior opinion and the opinions of its neighbors at time $t - 1$. Specifically,

$$x_i^v(t) = (1 - \alpha) \cdot x_i^v(t-1) + \alpha \sum_{u \in N^+(v)} w(u, v) \cdot x_i^u(t-1), \qquad \forall i \in M, \qquad (2)$$

where $N^+(v) = \{u \mid (u, v) \in E\}$ is the set of v's incoming neighbors in G and $0 < \alpha < 1$ is a model parameter.

This dynamic can be expressed more succinctly in matrix form. Let A be the weighted adjacency matrix of G with entries $a_{ij} = w(i, j)$, and define $\Gamma = (1 - \alpha)I + \alpha A^\mathsf{T}$ to be the $n \times n$ influence matrix[1] with entries γ_{vu} conveying the amount of direct influence that the opinions of node u shape those of node v from one time step to the next. By construction, Γ is an aperiodic stochastic matrix. Let $\mathbf{x}_i(t)$ denote the length-n vector containing entries for each node's opinion

---

[1] Γ is sometimes referred to as a *listening structure* [7] or *interaction matrix* [10].

toward player i at time t. Using Γ and $\mathbf{x}_i(\cdot)$, we can rewrite the dynamics in Eq. (2) as $\mathbf{x}_i(t) = \Gamma \mathbf{x}_i(t-1)$. In particular, the opinions toward player i after T time steps is simply $\mathbf{x}_i(T) = \Gamma^T \mathbf{x}_i(0)$.[2]

Upon the termination of the opinion dynamics after T steps, each node v is left holding an opinion vector $\mathbf{x}^v(T)$. The utility for player $i \in M$ is defined to be the average relative opinion held by the population toward i,

$$\pi_i(s) = \frac{1}{n} \sum_{v \in V} \frac{x_i^v(T)}{|\mathbf{x}^v(T)|_1}. \tag{3}$$

Notice that, (3) implies that for any strategy profile s, we have $\sum_{i \in M} \pi_i(s) = 1$, so the NIG is a constant-sum game.

2.1 Influence

A player's best response strategy in the NIG involves selecting an "influential" subset of seed nodes so that, upon the termination of the dynamics, the average relative opinion held by the nodes toward the player is maximized. The precise character of *influence* in this context deserves some examination.

It is well-known that the DeGroot consensus dynamic converges so that a common opinion is shared by every node in the network is guaranteed as $T \to \infty$ when the matrix Γ is aperiodic and stochastic. The consensus obtained is described by a weighted sum of the nodes' initial opinions (at time $t = 0$), with the weights given by the entries in the eigenvector of Γ corresponding to the eigenvalue 1. Hence, for sufficiently large T, a node's influence is directly related to their eigenvector centrality. However, if T is not large enough to obtain consensus, then a node's eigenvector centrality no longer corresponds to the influence they exert on the average opinions upon the termination of the dynamics.

The importance of T in characterizing the influence that a node exerts on a diffusion process was recently identified in an empirical study by Banerjee *et al.* [2] (see also [3]), which led them to define a quantity called *diffusion centrality*. Although their definition corresponds to a different diffusion dynamic than the one we consider in this paper, we can still offer a definition that is qualitatively similar to theirs but tailored to the consensus dynamic.

Let $\delta[j]$ denote the n-dimensional column vector consisting of a one in row j and zeros everywhere else. The diffusion centrality of a node v is defined to be the vector $\mathbf{c}^v = \Gamma^T \delta[v]$ whose entries c_u^v describe the fraction of node u's opinion at time T that is due to node v's initial opinion at time $t = 0$. As $T \to \infty$, the convergence of the consensus dynamic ensures that $|c_u^v - c_{u'}^v| \to 0$ for all $u, u' \in V$; let c^v denote this uniform amount of influence that v's brings to bear upon the final opinion of every node in V. The value c^v is precisely the v'th entry of the unique left eigenvector $\mathbf{c}$ of Γ corresponding to the eigenvalue 1, and the sum of the entries in $\mathbf{c}$ equal 1.

---

[2] We use Γ^T to denote the matrix Γ raised to the Tth power. For matrix transposition, we use the notation Γ^T.

3 On the Existence of Pure Nash Equilibrium

This section presents our results on the existence of pure Nash equilibrium strategies. Recall that a pure strategy profile $s = (s_i, s_{-i})$ is a Nash equilibrium if, for every player $i \in M$ and every possible pure strategy s'_i for that player, we have $\pi_i(s_i, s_{-i}) \geq \pi_i(s'_i, s_{-i})$. Throughout this paper, all of the results regarding Nash equilibrium will be with respect to pure strategies, and all unqualified mentions of *strategy* should be understood to refer to a *pure strategy*; all mentions of *Nash equilibrium* refer to *pure Nash equilibrium*.

3.1 At Consensus

In this section we establish the existence of pure strategy Nash equilibria for NIGs in which T is sufficiently large to ensure that the opinions reach a consensus. In the consensus regime, all nodes share the same final opinion vector, $\mathbf{x}(T) = (x_1(T), \ldots, x_m(T))$, and the utility for each player $i \in M$ is simply

$$\pi_i(s) = \frac{x_i(T)}{|\mathbf{x}(T)|_1}. \tag{4}$$

With the profile of weights $\mathbf{c} = (c^{v_1}, c^{v_2}, \ldots, c^{v_n})$, where c^v denotes the weight of node v's contribution to this consensus, we can express the consensus opinion toward player i as $x_i(T) = \sum_{v \in V} c^v x_i^v(0)$. Notice that the consensus opinion $x_i(T)$ for player i is a countably additive function of player i's strategy, s_i. This implies that in order for a deviation from s_i to s'_i to increase the consensus opinion toward player i, then any deviation from s_i to s''_i where $s''_i = (s_i \setminus \{v\}) \cup \{u\}$, $v \in s_i \setminus s'_i$, and $u \in s'_i \setminus s_i$ will also increase the consensus opinion toward player i. In other words, if i can increase their share of the consensus opinion by swapping k nodes in their strategy, then they can also gain from swapping only a single node.

A key feature to the consensus case is that, since all nodes will converge to the same opinion, there is no longer any need for players to make strategic trade-offs involving increasing their opinion share among one particular subset of nodes at the cost of decreasing their opinion share elswhere. This is reflected by the fact that the utility function in Eq. (3) reduces to Eq. (4) in the consensus regime (*i.e.*, when T is sufficiently large).

Proposition 1. *Every NIG in which T is sufficiently large so as to ensure that a consensus opinion $\mathbf{x}^v(T)$ is reached among every node $v \in V$ has at least one pure Nash equilibrium.*

Proof (Sketch). Assume that the m players are ordered so that $b_1 \geq b_2 \geq \cdots \geq b_m$ and the nodes $V = \{v_1, v_2, \ldots, v_n\}$ are ordered so that $c^{v_1} \geq c^{v_2} \geq \cdots \geq c^{v_n}$. Set player 1's seed strategy s_1 such that it maximizes $x_1(T)$. The strategies for players $i = 2, 3, \ldots, m$ will be built sequentially, so that s_i is a best response to the profile $s^{i-1} = (s_1, \ldots, s_{i-1})$. We then argue that, given $s^i = (s_1, \ldots, s_{i-1}, s_i)$, for every player $j < i$, s_j is a best response to $s^i_{-j} = (s_1, \ldots, s_{j-1}, s_{j+1}, \ldots, s_i)$.

Here, we will give the proof for $m = 2$ players; the proof for $m \geq 2$ players follows from an inductive argument that is based on similar reasoning.

Set $s_1 = \{v_1, \ldots, v_{b_1}\}$, and let s_2 be a best response to s_1 that maximizes the value of the consensus opinion toward player 2, $x_2(T)$. We have two cases to consider: (i) $s_1 \cap s_2 = \emptyset$, and (ii) $s_1 \cap s_2 \neq \emptyset$.

The first case is trivial: if s_2 does not contain any of the nodes in s_1, then player 1 enjoys exclusive access to b_1 of the most influential seed nodes.

For the second case, let $r = s_1 \cap s_2$ be the set of seed nodes chosen by player 2 that are also in s_1. Suppose, toward a contradiction, that player 1 can strictly benefit from changing to a strategy $s_1' = (s_1 \setminus \{u\}) \cup \{v\}$ for some nodes $u \in r$ and $v \in V \setminus (s_1 \cup s_2)$; i.e., player 1 swaps out a shared node u for exclusive access to another node v. By swapping out u for v, player 1 may increase the consensus opinion toward them self, but they would also be increasing the consensus opinion toward player 2 by relinquishing their share of the influence weight c^u. But, by virtue of player 2's inclusion of u instead of v in their own best response, it must be the case that the relative opinion toward player 1 – and, thus, player 1's utility – would not improve by swapping u for v. Therefore, $\pi_1(s_1', s_2) \leq \pi_1(s_1, s_2)$, contradicting the claim that s_1' can earn player 1 a strictly higher utility than s_1.   □

3.2   The General Case

In this section, we show that the existence of a pure Nash equilibrium is *not* guaranteed in NIGs when the dynamic does not reach a consensus opinion. We prove the non-existence for the symmetric setting, in which every player shares the same seed budget; a proof for the asymmetric setting can be found in the full version of this paper.

Proposition 2. *For any $m \geq 2$ and symmetric seed budget b, there exist NIGs that do not admit a pure strategy Nash equilibrium.*

Proof. A construction that does not admit a pure Nash equilibrium is as follows: Create $\mu = m(b+1)+1$ nodes $v_0, v_1, \ldots, v_{\mu-1}$, and add directed edges from each v_i to v_{i+k} for $k = 1, \ldots, b$. Next, for each v_i, add two additional "petal nodes" $v_{i,l}$ and $v_{i,r}$, and add the following four links: $(v_i, v_{i,l})$, $(v_i, v_{i,r})$, $(v_{i,l}, v_{i,r})$, and $(v_{i,r}, v_i)$. Define the weight of each edge (u, v) to be $w(u, v) = \frac{1}{|N^+(v)|}$. Note that the sum of each node's incoming edge weights equals one, and that the resulting graph is strongly connected. (See Fig. 1 for an example.)

Let G be a graph that implements the above construction, and set the parameters $\alpha = 1/2$ and $T = b$. Because a player's optimal strategy will never involve seeding a petal node, we restrict our attention to strategies comprising only central nodes. Let V_C denote these μ central nodes. Since $|V_C| = \mu = m(b+1) + 1$ and the total seed budget among all m players is bm, it will always be the case that a player's best response will not include nodes that are already included in another player's strategy, and exactly $m + 1$ of the central nodes will not be

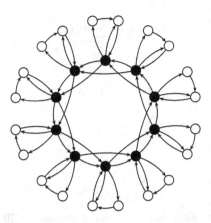

Fig. 1. Example of the graph construction in the proof of Proposition 2 with $m = 3$ players and budgets $b = 2$. The dark nodes are the "central nodes" and the white ones are the "petal nodes". This network does not admit a pure Nash equilibrium.

included as seed nodes in any players' strategies. Let $U \subset V_C$ denote these $m + 1$ unseeded central nodes. By design, the graph is constructed so that players prefer selecting seed nodes with successors in U. Of the $|U| = m + 1$ nodes, at most m of them will have no predecessors that are also in U; and there will always be at least one node $u^* \in U$ that does have a predecessor in U. Let u' denote u^*'s predecessor in U. A player i with a seed node $v \in s_i$ that is a successor of u' can always improve their utility by changing to a strategy $s_i' = \{u'\} \cup (s_i \setminus \{v\})$. Since the existence of (at least one) such u^* is guaranteed, then there will always be a player i that can change strategies for an increase in utility. Hence, a Nash equilibrium cannot be obtained. □

4 Computational Properties of Best Response

This section considers the computational problem of finding, for a given player i and strategy profile s_{-i}, a strategy s_i that maximizes i's utility, $\pi_i(s_i, s_{-i})$. Such a strategy s_i is called a *best response* to the profile s_{-i} of strategies belonging to every other player $j \neq i$.

Our first result in this section establishes the computational complexity of the best response problem.

Proposition 3. *Finding a best response strategy for the NIG is NP-complete.*

Proof (Sketch). Hardness follows by reduction from SET COVER and completeness is due to the fact that the utility function can be computed in polynomial time. □

Next, we turn the problem of finding an approximate best response. It will be useful to adopt the following definition of the utility function, which is equivalent to Eq. (3):

$$\pi_i(s) = \frac{1}{n} \sum_{v \in V} g_v^i(s) \tag{5}$$

where

$$g_v^i(s) = \frac{f_v^i(s)}{f_v(s)}, \tag{6}$$

$$f_v^i(s) = \sum_{u \in s_i} \frac{c_v^u}{m_u(s)} + \sum_{u \in V \setminus s} \varepsilon c_v^u, \tag{7}$$

and

$$f_v(s) = \sum_{j \in M} f_v^j(s).$$

Here, we employ the notation $m_u(s)$ to denote the cardinality of the set $M_u(s) = \{j \mid u \in s_j\} \subseteq M$. The quantities c_v^u, which measure the amount of influence that node u exerts on node v after T time steps, was defined in Sect. 2.1.

Our main result for this section establishes that the utility function $\pi_i(\cdot)$ is submodular,[3] from which it follows from the classic result by Nemhauser, Wolsey, and Fisher [16] that a $(1 - 1/e) \approx 0.6321$ approximation can be computed using a greedy algorithm.

Proposition 4. *The utility function $\pi_i(\cdot)$ is monotonic and submodular.*

Proof (Sketch). The submodularity and monotonicity follow immediately from establishing that (7) is increasing and (6) is submodular and the fact that since $\pi_i(\cdot)$ is a function that is defined by a linear combination of submodular functions (*cf.*, Eq. (5)), then $\pi_i(\cdot)$ is itself submodular.   □

5   Discussion and Future Work

This paper presented a model for the strategic seeding of opinion dynamics using the simple, well-studied DeGroot consensus dynamic. We established that the existence of pure Nash equilibria cannot be guaranteed if the dynamic is not allowed to run to consensus. The amount of time required for the dynamic to reach to consensus is known to be slower in networks with modular (homophilic) connectivity patterns [9]. This implies that in practice, strategic opinion seeding on real-world social networks, which often exhibit modular structures, should not assume that there will be enough time for the population to coalesce around a shared, consensus opinion; and, crucially, the individuals that appear to be attractive seeds in the steady state regime when a consensus is reached (those with high eigenvector centrality) may not be the best choice if the dynamics halt in the transient regime.

Our findings in Sect. 4 on the computational problem of finding best response strategies are in alignment with similar competitive influence models that

---

[3] A set function $f : \Omega \to \mathbb{R}$ is *submodular* if, for every $X \subseteq Y \subset \Omega$ and element $x \in \Omega \setminus Y$, we have $f(X \cup \{x\}) - f(X) \geq f(Y \cup \{x\}) - f(Y)$.

employ opinion dynamics that are more complex and less amenable to analytical tractability than the simplistic DeGroot consensus dynamic we use. For example, in a competitive extension of the probabilistic *independent cascade* model of Kempe *et al.* [14], Bharathi *et al.* [4] establish a $(1 - 1/e)$ approximation guarantee using monotone and submodularity arguments that extend those used in [14] for the "single-player" setting (see also Mossel and Roch [15]). However, some models have approximation guarantees that can be significantly worse than $(1 - 1/e)$. For example, Borodin *et al.* [5] show that competitive extensions of the *threshold* diffusion model do not share the $(1 - 1/e)$ approximation guarantee established in [14] for the "single player" optimization setting. They do, however, offer a variant of the threshold model that does admit a $(1 - 1/e)$ approximation guarantee. Similar $(1 - 1/e)$ approximation guarantees are established more recently in competitive influence maximization models by Goyal *et al.* [11] and Fotakis *et al.* [8].

Our analysis in Sect. 3 highlights the importance of the length of the diffusion process, T, in guaranteeing the existence pure strategy Nash equilibria. The identification of conditions that are sufficient to guarantee the existence of equilibria for small, non-consensus reaching values of T is an interesting open problem. Related to this is an intriguing extension to the model that would allow players to not only choose *which* nodes to seed, but also *when* to seed them. In our preliminary investigations into this extension, we have observed in simulations that some graphs contain nodes whose influence "peaks" at a greater magnitude in the dynamic's transient regime than in the steady state. We believe that such an extension would also more closely model many real-world applications, such as political contests and advertising campaigns, where timing can be an important consideration.

Acknowledgements. The authors gratefully acknowledge support from the US Army Research Office MURI Award No. W911NF-13-1-0340 and Cooperative Agreement No. W911NF-09-2-0053.

References

1. Alon, N., Feldman, M., Procaccia, A.D., Tennenholtz, M.: A note on competitive diffusion through social networks. Inf. Process. Lett. **110**(6), 221–225 (2010)
2. Banerjee, A., Chandrasekhar, A.G., Duflo, E., Jackson, M.O.: The diffusion of microfinance. Science **341**(6144), 363–371 (2013)
3. Banerjee, A.V., Chandrasekhar, A., Duflo, E., Jackson, M.O.: Gossip: Identifying central individuals in a social network. Working Paper, August 2014
4. Bharathi, S., Kempe, D., Salek, M.: Competitive influence maximization in social networks. In: Deng, X., Graham, F.C. (eds.) WINE 2007. LNCS, vol. 4858, pp. 306–311. Springer, Heidelberg (2007). doi:10.1007/978-3-540-77105-0_31
5. Borodin, A., Filmus, Y., Oren, J.: Threshold models for competitive influence in social networks. In: Saberi, A. (ed.) WINE 2010. LNCS, vol. 6484, pp. 539–550. Springer, Heidelberg (2010). doi:10.1007/978-3-642-17572-5_48
6. DeGroot, M.H.: Reaching a consensus. J. Am. Stat. Assoc. **69**(345), 118–121 (1974)

7. DeMarzo, P.M., Vayanos, D., Zweibel, J.: Persuasion bias, social influence, and unidimensional opinions. Q. J. Econ. **118**(3), 909–968 (2003)
8. Fotakis, D., Lykouris, T., Markakis, E., Obraztsova, S.: Influence maximization in switching-selection threshold models. In: Lavi, R. (ed.) SAGT 2014. LNCS, vol. 8768, pp. 122–133. Springer, Heidelberg (2014). doi:10.1007/978-3-662-44803-8_11
9. Golub, B., Jackson, M.O.: How homophily affects the speed of contagion, best response and learning dynamics. Working Paper, September 2010
10. Golub, B., Jackson, M.O.: Naïve learning in social networks and the wisdom of crowds. Am. Econ. J. Microeconomics **2**(1), 112–149 (2010)
11. Goyal, S., Heidari, H., Kearns, M.: Competitive contagion in networks. Games and Economic Behavior (2014) (in press)
12. Jackson, M.O.: Social and Economic Networks. Princeton University Press, Princeton (2008)
13. Jackson, M.O.: An overview of social networks and economic applications. In: Benhabib, J., Bisin, A., Jackson, M.O. (eds.) Handbook of Social Economics, chap. 12, pp. 511–585. North Holland, San Diego (2011)
14. Kempe, D., Kleinberg, J., Tardos, É.: Maximizing the spread of influence through a social network. In: Proceedings of the Ninth ACM SIGKDD International Conference on Knowledge Discovery and Data Mining, KDD 2003, pp. 137–146. ACM, New York (2003)
15. Mossel, E., Roch, S.: Submodularity of influence in social networks: from local to global. SIAM J. Comput. **39**(6), 2176–2188 (2010)
16. Nemhauser, G.L., Wolsey, L.A., Fisher, M.L.: An analysis of approximations for maximizing submodular set functions—i. Math. Program. **14**(1), 265–294 (1978)
17. Tzoumas, V., Amanatidis, C., Markakis, E.: A game-theoretic analysis of a competitive diffusion process over social networks. In: Goldberg, P.W. (ed.) WINE 2012. LNCS, vol. 7695, pp. 1–14. Springer, Heidelberg (2012). doi:10.1007/978-3-642-35311-6_1

Data Preservation in Base Station-Less Sensor Networks: A Game Theoretic Approach

Yutian Chen[1]([⊠]) and Bin Tang[2]

[1] Economics Department, California State University Long Beach, Long Beach, USA
Yutian.Chen@csulb.edu
[2] Computer Science Department, California State University Dominguez Hills,
Carson, USA
btang@csudh.edu

Abstract. We aim to preserve the large amount of data generated inside *base station-less sensor networks* with minimum energy cost, while considering that sensor nodes are selfish. Previous research assumed that all the sensor nodes are cooperative and designed a centralized minimum-cost flow solution. However, in a distributed setting wherein energy- and storage-constrained sensor nodes are under different control, they could behave selfishly, only to maximize their own benefit. In this paper, we take a game theoretic approach and design a computationally efficient data preservation game. We show that in our game, individual sensor nodes, motivated solely by self-interest, achieve good system-wide data preservation solution.

Keywords: Sensor networks · Data preservation · Energy-efficiency · Game theory

1 Introduction

Sensor networks are ad hoc multi-hop wireless networks formed by a large number of low-cost sensor nodes with limited battery power, storage spaces, and processing capacity. Wireless sensor networks have been used in a wide range of applications such as military surveillance, environmental monitoring, and target tracking [22]. Recently, some of the emerging sensor networks are deployed in challenging environments such as in remote or inhospitable regions, or under extreme weather, to continuously collect large volumes of data for a long period of time. Such emerging sensor networks include seismic sensor networks [5], underwater or ocean sensor networks [10,17,21], wind and solar harvesting [3,11], and volcano eruption monitoring and glacial melting monitoring [12,20].

In the above scenarios, it is not practical to deploy data-collecting base stations with power outlets in or near such inaccessible sensor fields. Due to the absence of the base stations, these sensor networks are referred to as *base station-less sensor networks*. Sensory data generated therefore have to be stored inside the network for some unpredictable period of time and then being collected by

© ICST Institute for Computer Sciences, Social Informatics and Telecommunications Engineering 2017
J. Cheng et al. (Eds.): GameNets 2016, LNICST 174, pp. 13–23, 2017.
DOI: 10.1007/978-3-319-47509-7_2

periodic visits of robots or data mules [16], or by low rate satellite link [6]. In particular, some sensor nodes are close to the events of interest and are constantly generating sensory data, depleting their own storage spaces. We refer to the sensor nodes with depleted storage spaces while still generating data as *source nodes*. The newly generated data that can no longer be stored at source nodes is called *overflow data*. To avoid data loss, overflow data is offloaded to sensor nodes with available storages (referred to as *storage nodes*). We call this process *data preservation in base station-less sensor networks*.

Since wireless communication consumes most of the battery power of sensor node, the key challenge is how to conserve sensors' battery power by minimizing the total energy consumption in data preservation. Tang et al. showed that this problem is equivalent to minimum cost flow problem [18], which can be solved optimally and efficiently [2]. Two fundamental assumptions are needed for the optimal data preservation algorithm in Tang et al. to work. First, the work assumes that all the storage nodes are selfless in the sense that they are willing to contribute their battery power and storage spaces to help offloading and storing the overflow data from the source nodes. Second, the optimal algorithm depends on full observability of the data preservation costs of each storage node, including the cost of relaying and storing the overflow data.

In this work, we tackle the data preservation problem when sensor nodes are selfish and are in lack of incentive to contribute to data preservation. Two reasons make it important to view sensor nodes as selfish players. First, sensor nodes are generally resource-constrained, with very limited amount of hardware resources including battery power, storage capacity, and processing power. Such resource constraints give sensor nodes minimum or zero motivation to be an altruistic player in data preservation. Second, in a large scale distributed sensor networks, sensor nodes could be under the control of different users or controllers, each of which pursues their own self-interest in the network. Under above scenarios sensor nodes can behave selfishly only to maximize their own benefit.

When sensor nodes are selfish, those assumptions in Tang et al. are no longer valid. First, in order to conserve their own battery power and storage spaces, the storage nodes will choose not to spend their energy and storage resources to help the source nodes to preserve the overflow data, obstructing the entire data preservation process. Second, the associated costs of data preservation of each storage node are normally private information which are not directly observed by outsiders. Due to selfishness, the storage nodes are in lack of incentive to truthfully report their costs. The reason is that each storage node needs to be paid in order to be motivated to participate in data preservation. Nonetheless through lying about its associated cost of data preservation, the storage node may successfully induce a data preservation path which generates itself a higher payoff compared to the payoff when it tells the truth. Such lying behavior of the storage nodes out of their selfishness clearly makes the data preservation path in Tang et al. inefficient. Therefore, with selfish sensor nodes, the challenge is to achieve good system performance, i.e., efficient data preservation with minimum energy cost, while still accommodating selfishness of the sensor nodes.

In this paper, we address the above challenge by utilizing the technique of **algorithmic mechanism design (AMD)** [13–15], a subfield of microeconomics and game theory. The goal of AMD is desirable – it designs computationally efficient game (including strategies and payoffs) such that individual players, motivated solely by self-interest, achieve good system-wide solution. We design computationally efficient data preservation game, in which a payment model is presented to compensate selfish nodes for participating in the data preservation. The payment to each node is designed in a way such that the following two purposes are achieved: first, each node, understanding how the payments are calculated, finds it optimal truthfully reporting its private cost information. Second, based on the reported cost of each node, the payment can sufficiently motivate each node who is involved in the optimal data preservation path calculated in Tang et al. to actually participate in data preservation. With these two goals achieved, the payment model in our game leads to good system-wide data preservation solution with each sensor node motivated solely by self-interest.

2   Data Preservation Problem

Network Model. The sensor network is represented as an undirected connected graph $G(V, E)$, where $V = \{1, 2, ..., n\}$ is the set of n sensor nodes and E is the set of m edges. The sensory data are modeled as a sequence of data packets, each of which is a bits. Some sensor nodes are close to the event of interest and generate large amount of data packets and deplete their storage spaces; they are referred to as *source nodes*. WLOG there are k source nodes $V_s = \{1, 2, ..., k\}$. The rest nodes in $V - V_s$ are referred to as *storage nodes*. Let d_i denote the number of overflow data packets source node i generates. Because of the storage depletion of the source nodes, the overflow data packets must be offloaded from their source nodes to some storage nodes to be preserved. Let $d = \sum_{i=1}^{k} d_i$ be the total number of overflow data packets, and let $D = \{D_1, D_2, ..., D_d\}$ denote the set of these d data packets. Let $s(j) \in V_s$, $1 \le j \le d$, denote D_j's source node. Let m_i be the available free storage space (in bits) at sensor node $i \in V$. If $i \in V_s$, then $m_i = 0$, implying that a source node is storage-depleted and thus has zero available storage space. If $i \in V - V_s$, then $m_i \ge 0$, implying that a storage node i can store another m_i bits of data packets. We assume that $\sum_{i=k+1}^{n} m_i \ge d \cdot a$, that is, the total size of the overflow data packets can be accommodated by the total available storage spaces.

<u>Energy Model.</u> We consider three different kinds of energy consumptions incurred in data preservation.

- *Transmitting Energy $E_i^t(j)$.* When node i sends a data packet of a bits to its one-hop neighbor j over their distance $l_{i,j}$, the amount of *transmitting energy* spent by i is $E_i^t(j) = a \cdot \epsilon_i^a \cdot l_{i,j}^2 + a \cdot \epsilon_i^e$. Here, ϵ_i^a is energy consumption of sending one bit on transmit amplifier of node i, and ϵ_i^e is energy consumption of transmitting one bit on the circuit of node i.

– *Receiving Energy* E_i^r. When node i receives an a-bit data packet from one of its one-hop neighbor, the amount of *receiving energy* it spends is $E_i^r = a \cdot \epsilon_i^e$. Here, ϵ_i^e is energy consumption of receiving one bit on the circuit of node i. Note that E_i^r does not depend on the distance between nodes.
– *Storing Energy* E_i^s. When node i stores a-bit data into its local storage, the amount of *storing energy* it consumes is $E_i^s = a \cdot \epsilon_i^s$. Here ϵ_i^s is the energy consumption of storing one bit at node i.

Problem Formulation. Define a *preservation function* as $p : D \rightarrow V - V_s$, indicating that a data packet $D_j \in D$ is offloaded from its source node $s(j) \in V_s$ to a storage node $p(j) \in V - V_s$ to be preserved. Let $P_j = \{s(j), ..., p(j)\}$ be the *preservation path* along which D_j is offloaded. Let $c_{i,j}$ denote node i's energy consumption in preserving D_j. $c_{i,j}$ can be represented as Eq. 1 below, with $\sigma(i, j)$ being the successor node of i on P_j.

$$c_{i,j} = \begin{cases} E_i^t(\sigma(i,j)) & i = s(j) \\ E_i^r + E_i^s & i = p(j) \\ E_i^r + E_i^t(\sigma(i,j)) & i \in P_j - \{s(j), p(j)\} \\ 0 & \text{otherwise} \end{cases} \qquad (1)$$

The objective is to find a preservation function p and P_j ($1 \le j \le d$) to minimize the *total preservation cost*, denoted as c, i.e.,

$$c = \min_p \sum_{j=1}^{d} \sum_{i=1}^{n} c_{i,j} = \min_p \sum_{i=1}^{n} \sum_{j=1}^{d} c_{i,j}, \qquad (2)$$

under the storage constraint that the total size of data offloaded to storage node i can not exceed i's storage capacity: $|j| 1 \le j \le d, p(j) = i| \cdot a \le m_i, \forall i \in V - V_s$.

Algorithm. Tang et al. [18] has shown that this problem is equivalent to the minimum cost flow problem in a properly transformed graph of the sensor network graph. The minimum cost flow problem can be solved optimally and efficiently [2]. We adopt and implement the scaling push-relabel algorithm proposed in [1,7]. It has the time complexity of $O(|V|^2|E|\log(|V|C))$, where C is the maximum capacity of an edge in the transformed graph. We denote the algorithm designed in Tang et al. [18] as *the centralized algorithm* to highlight that it minimizes data preservation energy based on the assumption that each node in the network is selfless and therefore fully cooperative.

In this work, we instead consider selfishness of nodes in the sense that each node is maximizing its own interest instead of the system interest. The central problem is to design a mechanism to incentivize selfish nodes to accomplish data preservation as in the centralized algorithm. Note that each source node is obligated to offload its data therefore selfishness does not apply to source nodes. On the other hand, storage nodes are selfish and need to be motivated. However, selfishness of storage nodes can lead to two problems. First, each storage node has no incentive to either relay or store data as either task consumes energy. Therefore, our mechanism needs to pay those storage nodes involved in data preservation path solved from the centralized algorithm, in order to give them incentive to participate in data preservation. The second problem is more subtle but fundamental. The centralized algorithm can figure out the minimum cost data preservation path only based on the assumption that data preservation costs of each storage node are observed. However, some of those cost parameters of each node (given by ϵ_i^e, ϵ_i^a and ϵ_i^s) are private information of each node and may not be directly observed by outsiders. Thus our mechanism needs to induce each node to truthfully report their unobserved cost parameters, so that the centralized algorithm can calculate the minimum cost path based on the reported cost parameters.

3   Algorithmic Mechanism Design (AMD) Approach

The goal of AMD is to design a game in which selfish players maximizing their own utility will choose strategies resulting in the social optimum specified by an optimal algorithm. Here the resulted state is referred to as the *dominant strategy equilibrium/solution*. Dominant strategy of a player is a strategy always maximizing his utility regardless of the other players' strategies. In a dominant strategy solution, each player is playing his dominant strategy. Note that a dominant strategy solution is also a Nash equilibrium since no player has an incentive to deviate from its strategy unilaterally. The challenge in the data preservation problem is to design utility function so that truthfully reporting its cost parameter is a *dominant strategy* to each storage node. Below we first introduce the concepts and notations of the AMD model. We then present the payment model, and prove that under this payment model, acting truthfully (that is, telling its true energy cost involved in data preservation) is each node's dominant strategy.

The AMD Model. There are n nodes in the network - node i has some private information t_i, called its type. There is an *output specification* that maps each type vector $t = \{t_1, ..., t_n\}$ to some output o. Node i's cost is given by *valuation function* $v_i(t_i, o)$, which depends on t_i as well as o. A *mechanism* defines for each node i is a set of strategies A_i. When i plays strategy $a_i \in A_i$, the mechanism computes an *output* $o = o(a_1, ..., a_n)$ and a *payment vector* $p = (p_1, ..., p_n)$, where $p_i = p_i(a_1, ..., a_n)$. Node i wants to maximize its utility function $\pi_i(a_1, .., a_n) = v_i(t_i, o) + p_i$.

There are three observations of the total preservation cost (Eq. 2). First, the total cost is the sum of all participating nodes' energy costs. We can therefore adopt the Vickrey-Groves-Clark (VCG) mechanism [4,9,19]. VCG mechanism applies to mechanism design optimization problems where the objective function is simply the sum of all agents' valuations, and it guarantees that each agent plays truthfully by reporting its true valuation [13]. Second, to minimize the total preservation cost for all the data packets, it only needs to minimize the preservation cost for each data packet D_j, given its source node and destination node. Therefore, our payment model focuses on only one packet, say D_j. Third, in the context of data preservation, $c_{i,j}$ is essentially the private information held by storage node i. To each storage node i, its strategy set includes to truthfully report its cost parameter (therefore $c_{i,j}$) or to lie about its cost parameter. That is, $(c_{i,1}, c_{i,2}, ..., c_{i,d}) \in A_i$ and $v_i(t_i, o) = -c_{i,j}$ for any i. Therefore i's utility is $\pi_i = p_i - c_{i,j}$.

Payment and Utility Model. Below we present the payment and utility model. Since we focus on any data packet D_j (and its preservation path P_j), we use c_i instead of $c_{i,j}$ to denote node i's true cost. p_i is the payment made to node i in order to motivate it to participate the data preservation, $\pi_i = p_i - c_i$. Let c_{-i} denote the strategies of all other nodes except node i.

Definition 1 Payment and Utility. *Based on Green and Laffont [8], under VCG mechanism, given any cost $\tilde{c}_i$ reported by node i, the amount of payment given to node i depends on whether node i is chosen to participate in data preservation according to the centralized algorithm. Its payment is 0 if it is not chosen; and its payment when it is chosen is:*

$$p_i(\tilde{c}_i, c_{-i}) = c_{V-\{i\}} - (\tilde{c}_V - \tilde{c}_i), \qquad (3)$$

where $c_{V-\{i\}}$ is the minimum total cost of the preservation path that does not go through i; $\tilde{c}_V$ is the minimum total cost of the preservation path that goes through i, when i reports its cost $\tilde{c}_i$. Therefore i's utility is 0 when it is not chosen by the centralized algorithm; and when i is chosen, its utility is

$$\pi_i(\tilde{c}_i, c_{-i}) = p_i(\tilde{c}_i, c_{-i}) - c_i = c_{V-\{i\}} - (\tilde{c}_V - \tilde{c}_i) - c_i, \qquad (4)$$

where c_i is node i's true cost. Moreover, we define c_V as the minimum total cost of the preservation path that goes through i when i truthfully reports its cost, i.e., when $\tilde{c}_i = c_i$.   □

Time complexity of the payment model. The time taken to compute the payment model is the time taken for the minimum cost flow calculation, which is $O(|V|^2|E|\log(|V|C))$, where C is the maximum capacity of an edge in the transformed graph [1,7]. Under this model, the amount of payment given to a specific node i equals the total minimum cost of all the participating nodes when i does not participate minus all other participating nodes' cost when i participates. The rationale is that a node can be motivated to participate if it is paid its share of contribution, which in our case, is the amount of preservation energy this node helps to reduce when it participates.

An implication here is that the payment and utility model is common knowledge to each node. That is, each node understands that based on their reported cost types and the corresponding data preservation path calculated by the centralized algorithm, their payment and utility are given by (3) and (4), respectively. The timing of the game among the source nodes is given below.

Definition 2 Timing of the Game. *The game unfolds as follows. In stage 1, each storage node reports its private type c_i. In stage 2, the centralized algorithm is applied based on reported cost types to calculate the minimum cost data preservation path. In stage 3, each of the storage nodes chosen in the path chooses to* *participate in data preservation or not. If they participate, they realize the data preservation cost and also the payment given by Eq. (3), and each gets utility given by Eq. (4).*   □

Note that each storage node moves only in stages 1 and 3, when each chooses how much to report for private type and whether to participate in data preservation based on the corresponding payment. Stage 2 is non-strategic: in the absence of base stations, the centralized algorithm is provided by an outsider of the system, and it cannot be enforced in the system by the outsider. Since there is a time sequence between the two decisions of each node in stage 1 and stage 3, the solution concept of the game is subgame perfect Nash equilibrium (SPNE). SPNE is a Nash equilibrium (NE) in which players are doing NE in every subgame of the whole game tree.

Assumptions. We assume that the source nodes are obliged to offload their overflow data packets to other storage nodes, thus need not to be motivated. Therefore, their types are known public knowledge, and they will be reimbursed according to true costs they entail. For storage node i that participates in the preservation of a specific data packet, it incurs one of the two costs below:

- *Relaying Cost* $c_i^r(j)$. When node i receives a data packet and then sends it to one of its one-hop neighbor j over their distance $l_{i,j}$, its *relaying cost*, denoted as $c_i^r(j)$, is the sum of its receiving energy and transmitting energy. That is $c_i^r(j) = E_i^r + E_i^t(j) = 2 \cdot a \cdot \epsilon_i^e + a \cdot \epsilon_i^a \cdot l_{i,j}^2$.
- *Storing Cost* c_i^s. When node i receives a data packet and then stores it into its storage, its *storing cost*, denoted as c_i^s, is the sum of its receiving energy and its storing energy. That is, $c_i^s = a \cdot \epsilon_i^e + a \cdot \epsilon_i^s$.

Note that node i has three energy parameters: ϵ_i^e, ϵ_i^a, and ϵ_i^s. Among them, ϵ_i^e affects both $c_i^r(j)$ and c_i^s, while ϵ_i^a only affects $c_i^r(j)$ and ϵ_i^s only affects c_i^s. Next, we will study the AMD model wherein for each node i, either ϵ_i^a or ϵ_i^s or ϵ_i^e is the private type of node i not directly observed by the public. Since ϵ_i^a and ϵ_i^s each only affects one cost or the other, we study $t_i = \epsilon_i^a$ or $t_i = \epsilon_i^s$ first.

3.1 AMD When $t_i = \epsilon_i^a$ or $t_i = \epsilon_i^s$

We focus on $t_i = \epsilon_i^a$ since $t_i = \epsilon_i^s$ can be studied similarly. Below we give a detailed proof that under above VCG payment model, for each node i, truth-telling (reporting its true type t_i) is a dominant strategy. We define $c_{-i}^r = \{c_1^r, \ldots, c_{i-1}^r, c_{i+1}^r, \ldots, c_n^r\}$ as the cost vector of other nodes except i. Since the optimal minimum cost flow algorithm determines i's successor node j, and for ease of notation, we use c_i instead of $c_i^r(j)$ to represent node i's relaying cost, and use $\tilde{c}_i$ instead of $\tilde{c}_i^r(j)$ in the following theorem and proof.

Theorem 1: *For any node i, suppose $t_i = \epsilon_i^a$ (that is, ϵ_i^a is i's private type). Reporting its true type ϵ_i^a is node i's dominant strategy. That is, $\pi_i(c_i, c_{-i}) \geq \pi_i(\tilde{c}_i, c_{-i})$, $\forall \tilde{c}_i \neq c_i$ and $\forall c_{-i}$.*

Proof: We consider that node i either reports truthfully or not. Under either case, node i could be chosen to participate in the data preservation or not according to the centralized algorithm. Therefore there are all together four cases. Below we show that $\pi_i(c_i, c_{-i})$ is always greater or equal to $\pi_i(\tilde{c}_i, c_{-i})$ in all the four cases.

Case I: Node i is in the preservation path when reporting either c_i or $\tilde{c}_i$. Thus the payment of i when it reports truthfully is $\pi_i(c_i, c_{-i}) = c_{V-\{i\}} - (c_V - c_i) - c_i = c_{V-\{i\}} - c_V$. On the other side, when it lies by reporting $\tilde{c}_i$, its payoff is $\pi_i(\tilde{c}_i, c_{-i}) = c_{V-\{i\}} - (\tilde{c}_V - \tilde{c}_i) - c_i = c_{V-\{i\}} - (\tilde{c}_V - \tilde{c}_i + c_i) = c_{V-\{i\}} - c_V$. Therefore in this case $\pi_i(c_i, c_{-i}) = \pi_i(\tilde{c}_i, c_{-i})$. Note that $\pi_i(c_i, c_{-i}) \geq 0$ because $c_{V-\{i\}} - c_V \geq 0$.

Case II: Node i is in the preservation path when reporting c_i, which implies that $c_{V-\{i\}} \geq c_V$; and it is not in the preservation path when reporting $\tilde{c}_i$, which gives payoff $\pi_i(\tilde{c}_i, c_{-i}) = 0$. Thus its payoff under truth-telling is $\pi_i(c_i, c_{-i}) = c_{V-\{i\}} - c_V \geq 0$. In this case $\pi_i(c_i, c_{-i}) \geq \pi_i(\tilde{c}_i, c_{-i})$.

Case III: Node i is not in the preservation path when reporting c_i, which gives $\pi_i(c_i, c_{-i}) = 0$ and also implies that $c_{V-\{i\}} \leq c_V$. However, it is in the preservation path when reporting $\tilde{c}_i$. Its payoff when it lies is $\pi_i(\tilde{c}_i, c_{-i}) = c_{V-\{i\}} - (\tilde{c}_V - \tilde{c}_i) - c_i = c_{V-\{i\}} - (\tilde{c}_V - \tilde{c}_i + c_i) = c_{V-\{i\}} - c_V \leq 0$. Therefore in this case $\pi_i(c_i, c_{-i}) \geq \pi_i(\tilde{c}_i, c_{-i})$.

Case IV: Node i is not in the preservation path when reporting either c_i or $\tilde{c}_i$. In this case $\pi_i(c_i, c_{-i}) = \pi_i(\tilde{c}_i, c_{-i}) = 0$.

Since $\pi_i(c_i, c_{-i}) \geq \pi_i(\tilde{c}_i, c_{-i})$ holds regardless of other nodes' strategy c_{-i} under all the cases, we conclude that reporting its true cost c_i is node i's dominating strategy.

Theorem 2. *With the payment given by (3), when ϵ_i^a or ϵ_i^s is unobserved, there exists SPNE of the game, in which every storage node i truthfully reports its cost type in stage 1. Moreover, in stage 3 each node i chosen by the centralized algorithm for data preservation will participate.*

Proof: In stage 3, each storage node chosen by the centralized algorithm will participate as long as its utility (i.e., payoff) is no less than zero. When we move back to stage 1, by Theorem 1, each storage node has a dominant strategy, which is to truthfully report the type. Therefore, the Nash equilibrium in stage 1 is that each node truthfully reports its type. Since each node in the data preservation path gets a non-negative utility (see the proof of Theorem 1), each will choose to participate in stage 3. We conclude that the strategy given in this theorem constitutes a SPNE. ∎

3.2 AMD When $t_i = \epsilon_e^i$

When ϵ_e^i is the unknown type of source node i, the reported value of the type affects the two costs simultaneously: the relaying cost $c_i^r(j)$ and the storing cost c_i^s. The complication here is that by lying about its type, node i might switch its role in data preservation from one task to a different task. For example, node

i might be assigned to relay the data packet according to ϵ_e^i, its true cost type; but by reporting $\tilde{\epsilon}_e^i \neq \epsilon_e^i$, node i might instead be assigned to store the data. It is not clear whether VCG can continue to apply in this case or not. To examine the situation when ϵ_e^i is the unknown type of node i, we first denote c_V^{is} as the minimum total cost of data preservation given that node i stores the data packet; and c_V^{ir} as the minimum total cost of data preservation given that node i relays the data packet. Note that $c_{V-\{i\}}$ is the minimum total cost of data preservation given that node i does not participate in data preservation. The following theorem shows that the basic idea of VCG continues to hold. Since the optimal minimum cost flow algorithm determines i's successor node j, and for ease of notation, we use c_i to represent $c_i^r(j)$ or c_i^s, and use $\tilde{c}_i$ to represent $\tilde{c}_i^r(j)$ or $\tilde{c}_i^s$ in the following theorem and proof.

Theorem 3. *For any node i, suppose $t_i = \epsilon_i^e$ (that is, ϵ_i^e is i's private type). Reporting its true type ϵ_i^e is node i's dominant strategy. That is, $\pi_i(c_i, c_{-i}) \geq \pi_i(\tilde{c}_i, c_{-i})$, $\forall\, \tilde{c}_i \neq c_i$ and $\forall\, c_{-i}$.*

Proof: Based on the reported ϵ_i^e of node i, node i could be chosen to participate in the data preservation or not according to the centralized algorithm. If it is chosen, it may be designated to either transmit or store the data packet. We need to show that regardless of other nodes' reported cost types, telling truth is always the optimal strategy of node i. There are in together six cases and we show that $\pi_i(c_i, c_{-i})$ is always greater or equal to $\pi_i(\tilde{c}_i, c_{-i})$ in all the six cases.

Case I. Node i is in the preservation path to relay the data packet when reporting either c_i or $\tilde{c}_i$.

Case II. Node i is in the preservation path to store the data packet when reporting either c_i or $\tilde{c}_i$.

Case III. Node i is in the preservation path to relay the data packet when reporting c_i and is doing nothing when reporting $\tilde{c}_i$.

Case IV. Node i is in the preservation path to store the data packet when reporting either c_i or $\tilde{c}_i$.

Proof for $\pi_i(c_i, c_{-i}) \geq \pi_i(\tilde{c}_i, c_{-i})$ for the four cases are similar as in the proof of Theorem 1 and are omitted. We focus on the following two cases.

Case V. Node i is in the preservation path to relay the data packet when reporting c_i and is in the preservation path to store the data packet when reporting $\tilde{c}_i$. This implies that $c_V^{is} \geq c_V^{ir}$. When node i reports c_i, its payoff is $c_{V-\{i\}} - c_V^{ir} + c_i - c_i = c_{V-\{i\}} - c_V^{ir}$. When node i reports $\tilde{c}_i$, its payoff is $c_{V-\{i\}} - \tilde{c}_V^{is} + \tilde{c}_i - c_i = c_{V-\{i\}} - c_V^{is}$. It follows that $c_{V-\{i\}} - c_V^{ir} \geq c_{V-\{i\}} - c_V^{is}$.

Case VI. Node i is in the preservation path to store the data packet when reporting c_i and is in the preservation path to relay the data packet when reporting $\tilde{c}_i$. This implies that $c_V^{is} \leq c_V^{ir}$. When node i reports c_i, its payoff is $c_{V-\{i\}} - c_V^{is} + c_i - c_i = c_{V-\{i\}} - c_V^{is}$. When node i reports $\tilde{c}_i$, its payoff is $c_{V-\{i\}} - \tilde{c}_V^{ir} + \tilde{c}_i - c_i = c_{V-\{i\}} - c_V^{ir}$. It follows that $c_{V-\{i\}} - c_V^{is} \geq c_{V-\{i\}} - c_V^{ir}$. ∎

Theorem 4: *With the payment given by (3), when ϵ_e^i is unobserved, there exists SPNE of the game, in which every source node i truthfully reports its cost type* *in stage 1. Moreover, in stage 3 each node i chosen by the centralized algorithm for data preservation will participate.*

Proof: It follows the same argument as the proof of Theorem 2 and is omitted here.   ∎

4   Conclusion and Future Work

In this work, we study data preservation problem in base station-less sensor networks wherein energy- and storage-constrained sensor nodes behave selfishly. We take a game theoretic approach and design a payment model under which the individual sensor nodes, motivated solely by self-interest, achieve good system-wide data preservation solution. In particular, we break down the data preservation cost of each storage node into two parts: relaying cost and storing cost, where cost parameters are node-dependent. The payment model is designed in a way such that no matter which cost parameter (related only to the relaying cost or only to the storing cost or to both) is private to the node, truthfully reporting the cost parameter is a dominant strategy to each node. We show that as a result, in the game it is an equilibrium that each storage node first truthfully reports its cost parameter, then participates in data preservation if it is chosen by the centralized data preservation algorithm.

In the next step of the work, we will validate theoretical findings using simulation results. By contrasting the payment of each storage node in the sensor network under truth-telling strategy to what it is under lying, we will show that truth-telling is never worse off and in certain cases is strictly better off to each storage node regardless of the choice of the other nodes. The simulation results thus can verify that truth-telling is a dominant strategy of each source node. Other future work includes relaxing some assumptions in the current work. In particular, we have assumed that data preservation is feasible in the sensor network, i.e., all the nodes have enough energy to offload and preserve all the overflow data packets. If instead the network is infeasible so that some data packets will inevitably be lost, it is interesting to see how the payment model can work to induce the efficient data preservation. Finally, we will extend our analysis to a dynamic scenario wherein overflow data are generated from time to time at different nodes. It is well understood in game theory that an infinitely repeated game gives a much larger set of equilibrium and in certain scenarios full cooperation can be achieved. In our setting of data preservation among selfish nodes, it is interesting to see to what extent we need to provide motivation for selfish storage node to engage in the optimal data preservation.

Acknowledgment. This work was supported in part by NSF Grant CNS-1116849.

References

1. Andrew goldberg's network optimization library. http://www.avglab.com/andrew/soft.html
2. Ahuja, R.K., Magnanti, T.L., Orlin, J.B.: Network Flows: Theory, Algorithms, and Applications. Prentice-Hall Inc., Englewood Cliffs (1993)
3. Cammarano, A., Petrioli, C., Spenza, D.: Pro-energy: a novel energy prediction model for solar and wind energy-harvesting wireless sensor networks. In: IEEE 9th International Conference on Mobile Adhoc and Sensor Systems (MASS 2012)
4. Clarke, E.H.: Multipart pricing of public goods. Public Choice (1971)
5. Cochran, E., Lawrence, J., Christensen, C., Chung, A.: A novel strong-motion seismic network for community participation in earthquake monitoring. IEEE Inst. Meas. **12**(6), 8–15 (2009)
6. Colitti, W., Steenhaut, K., Descouvemont, N., Dunkels, A.: Satellite based wireless sensor networks: Global scale sensing with nano- and pico-satellites. In: Proceedings of the 6th ACM Conference on Embedded Network Sensor Systems, SenSys 2008, pp. 445–446 (2008)
7. Goldberg, A.V.: An efficient implementation of a scaling minimum-cost flow algorithm. J. Algorithms **22**, 1–29 (1997)
8. Green, J., Laffont, J.: Incentives in public decision making. Stud. Public Econ. **1**, 65–78 (1979)
9. Groves, T.: Incentives in teams. Econometrica (1973)
10. Li, S., Liu, Y., Li, X.: Capacity of large scale wireless networks under gaussian channel model. In: Proceedings of MOBICOM 2008 (2008)
11. Li, Y., Li, X., Wang, P.: A module harvesting wind and solar energy for wireless sensor node. In: Wang, R., Xiao, F. (eds.) CWSN 2012. CCIS, vol. 334, pp. 217–224. Springer, Heidelberg (2013). doi:10.1007/978-3-642-36252-1_20
12. Martinez, K., Ong, R., Hart, J.K.: Glacsweb: a sensor network for hostile environments. In: Proceedings of SECON 2004 (2014)
13. Nisan, N.: Algorithms for selfish agents: mechanism design for distributed computation. In: Meinel, C., Tison, S. (eds.) STACS 1999. LNCS, vol. 1563, pp. 1–15. Springer, Heidelberg (1999). doi:10.1007/3-540-49116-3_1
14. Nisan, N., Ronen, A.: Algorithmic mechanism design. In: Proceedings of the Thirty-First Annual ACM Symposium on Theory of Computing (STOC 1999), pp. 129–140 (1999)
15. Nisan, N., Ronen, A.: Algorithmic mechanism design. Games Econ. Behav. **35**, 166–196 (2007)
16. Shah, R.C., Roy, S., Jain, S., Brunette, W.: Data mules: modeling a three-tier architecture for sparse sensor networks. In: Proceedings of SNPA 2003 (2003)
17. Syed, A.A., Ye, W., Heidemann, J.: T-lohi: a new class of mac protocols for underwater acoustic sensor networks. In: Proceedings of INFOCOM 2008 (2008). http://www.isi.edu/ilense/snuse/index.html
18. Tang, B., Jaggi, N., Wu, H., Kurkal, R.: Energy-efficient data redistribution in sensor networks. ACM Trans. Sen. Netw. **9**(2), 11: 1–11: 28 (2013)
19. Vickrey, W.: Counterspeculation, auctions and competitive sealed tenders. J. Finan. **16**, 8–37 (1961)
20. Werner-Allen, G., Lorincz, K., Johnson, J., Lees, J., Welsh, M.: Fidelity and yield in a volcano monitoring sensor network. In: Proceedings of OSDI 2006 (2006)
21. Xiao, Y.: Underwater Acoustic Sensor Networks. Auerbach Publications (2009)
22. Yick, J., Mukherjee, B., Ghosal, D.: Wireless sensor network survey. Comput. Netw. **52**, 2292–2330 (2008)

Energy Efficient Clustering and Beamforming for Cooperative Multicell Networks

Yawen Chen[1,2(✉)], Xiangming Wen[1,2], Zhaoming Lu[1,2], Hua Shao[1,2], Jingyu Lu[1,2], and WenPeng Jing[1,2]

[1] Beijing Key Laboratory of Network System Architecture and Convergence, Beijing University of Posts and Telecommunications, Beijing, China
{chenyw,xiangmw,sarathy,jingwenpeng}@bupt.edu.cn, lzy_0372@163.com, ljyab123@163.com
[2] Beijing Laboratory of Advanced Information Networks, Beijing, China

Abstract. Network densification is the most important way to improve the network capacity and hence is widely adopted to handle the ever-increasing mobile traffic demand. However, network densification will make the inter-cell interference severe and also significantly increase the energy budget. Multicell cooperative transmission is an efficient way to mitigate the inter-cell interference and plays an important role in energy efficiency optimization. This paper investigates the energy efficient multicell cooperation strategy for dense wireless networks. Joint cluster forming and beamforming are considered to optimize the energy efficiency (evaluated by bits/Hz/J). The optimization problem is then decoupled into two subproblems, i.e., energy efficient beamforming problem and energy efficient cluster forming problem. The fractional programming and Lagrangian duality theory are used to obtain the optimal beamformer. Coalition formation game theory is exploited to solve the cluster forming problem. The proposed energy efficient clustering and beamforming strategy can provide flexible network service according to spatially uneven traffic and greatly improve the network energy efficiency.

Keywords: Cooperative transmission · Energy efficiency · Beamforming · Clustering · Coalition formation game

1 Introduction

Network densification has been historically adopted for network capacity improvement and it will be persist in the future to handle the increasingly growing wireless traffic [16,18]. However, the dense deployment of base station (BS) leads to the network energy consumption increase. As the greatly increased energy efficiency has been listed as one of the main objectives when designing 5G wireless network [1], it is necessary to design more energy efficient strategies. Multicell cooperative transmission (MCT) is an efficient technique in interference managements [13,17]. The researches in [12] show that when all the BSs are coordinated, the spectral efficiency scales linearly with the signal-to-noise ratio

© ICST Institute for Computer Sciences, Social Informatics and Telecommunications Engineering 2017
J. Cheng et al. (Eds.): GameNets 2016, LNICST 174, pp. 24–33, 2017.
DOI: 10.1007/978-3-319-47509-7_3

(SNR). The basic idea of MCT is to let multiple BSs cooperate and act as a single Multiple Input Multiple Output (MIMO) transceiver, hence some of the interference are turned into useful signals. Consequently, the network throughput is greatly improved. The MCT also plays a vital role in energy efficiency optimization. In [5], an energy efficiency analysis framework for MCT is proposed and the results show that when the backhauling and cooperative processing power are carefully controlled, MCT can be energy efficient, especially for cell-edge communication.

The two main problems for energy efficient MCT design are how to form clusters with desirable size, and how to efficiently obtain the multicell beamformers. The gains of MCT are saturated with the growing number of cooperating BSs due to the excessive overheads, e.g., complexity, channel estimation and increased power consumption [8]. Hence the cooperative clusters have to be carefully designed to obtain the optimal trade-off between the performance gain and associated overhead. In [3], a novel affinity propagation model is used to semi-dynamically form the cooperative cluster. The proposed algorithm can greatly improve network throughput with low complexity. Coalition formation game (CFG) studies the complex interactions among players and the formation of cooperating groups, referred to as coalitions [11]. Hence, CFG is well suited for the BSs clustering problem and have been studied in [15], where CFG is used to form the small cells cluster to optimize the trade-off between the benefits and costs associated with cooperation. On the other hand, energy efficient beamforming is also important for MCT. In [4], the downlink-uplink duality theory and geometric programming are used to find the beamformer that maximize the network energy efficiency (EE, defined as the sum throughput to power consumption). In [14], minimum mean square error (MMSE) based energy efficient beamforming strategy is proposed to maximize the worst-case EE. However, [4,14] only focus on the static cooperative cluster.

In this paper, we study the energy efficient clustering and beamforming problem for cooperative multicell networks. A energy efficiency optimization problem that jointly considers dynamic clustering and beamforming is formulated. Due to the combinatorial nature of the clustering and beamforming in cooperative multicell networks, the joint optimization problem is extremely hard to solve. To efficiently solve it, the problem is decoupled into two subproblem, i.e., the cluster forming problem and the energy efficient beamforming problem. The CFG and fractional program are used to solve them respectively. The remainder of this article is organized as follows. We first describe the system model in Sect. 2. Then the energy efficient clustering and beamforming problem is formulated and efficiently solved in Sect. 3. In Sect. 4, numerical results are presented. Finally, we conclude the article in Sect. 5.

2   System Model

2.1   Signal Model

We consider the downlink of a cooperative multicell network where a set $\mathcal{Q}$ of BSs, each equipped with M antennas, is serving a set $\mathcal{I}$ of user equipments (UEs) equipped with N antennas. Assume that the BSs are partitioned into several clusters and jointly serve UEs. Let $Q = (Q_1, \ldots, Q_k, \ldots, Q_K)$ denote a partition of $\mathcal{Q}$ and Q_k denotes the kth cooperative cluster. The set of serving UEs in cluster k is denoted as I_k. Let $\mathbf{H}_{i_k}^{ql} \in \mathbb{C}^{N \times M}$ denote the channel between qth BS in cluster l and the ith user in the kth cluster. $\mathbf{H}_{i_k}^{l} \in \mathbb{C}^{N \times MQ_l}$ denotes the channel matrix between all BSs in cluster l to user i_k. Assume that each BS only transmit a single data stream to each UE and let $\mathbf{v}_{i_k}^{q_k} \in \mathbb{C}^{M \times 1}$ denote the beamformer from BSs q_k to UE i_k. Let $\mathbf{v}_{i_k} = \left[(\mathbf{v}_{i_k}^{1})^{H}, \ldots, (\mathbf{v}_{i_k}^{Q_k})^{H} \right]^{H} \in \mathbb{C}^{MQ_k \times 1}$ denote the beamformer collection intended for user i_k. Denote the transmitted signal for UE i_k as s_{i_k}, and then the received signal of user i_k can be expressed as

$$\mathbf{y}_{i_k} = \mathbf{H}_{i_k}^{k} \mathbf{v}_{i_k} s_{i_k} + \sum_{j_k \neq i_k} \mathbf{H}_{i_k}^{k} \mathbf{v}_{j_k} s_{j_k} + \sum_{l \neq k} \sum_{j_l \in I_l} \mathbf{H}_{i_k}^{l} \mathbf{v}_{j_l} s_{j_l} + \mathbf{z}_{i_k} \tag{1}$$

Let $\mathbf{u}_{i_k} \in \mathbb{C}^{N \times 1}$ denote the receiver beamformer to decode the intended signal. The estimated signal is $\hat{s}_{i_k} = \mathbf{u}_{i_k}^{H} \mathbf{y}_{i_k}$. Then the mean square error (MSE) of UE i_k can be calculated as

$$\begin{aligned} e_{i_k} &= \mathbb{E}_{s,\mathbf{z}} \left[(\hat{s}_{i_k} - s_{i_k})(\overline{\hat{s}_{i_k}} - \overline{s_{i_k}}) \right] \\ &= (1 - \mathbf{u}_{i_k}^{H} \mathbf{H}_{i_k}^{k} \mathbf{v}_{i_k})(1 - \overline{\mathbf{u}_{i_k}^{H} \mathbf{H}_{i_k}^{k} \mathbf{v}_{i_k}}) + \sum_{(l,j) \neq (k,i)} \mathbf{u}_{i_k}^{H} \mathbf{H}_{i_k}^{l} \mathbf{v}_{j_l} \mathbf{v}_{j_l}^{H} (\mathbf{H}_{i_k}^{l})^{H} \mathbf{u}_{i_k} + \sigma^2 \mathbf{u}_{i_k}^{H} \mathbf{u}_{i_k} \end{aligned}$$

$$\tag{2}$$

The achievable rate of UE i_k can be expressed as

$$R_{i_k} = \log \left| \mathbf{I}_N + \mathbf{H}_{i_k}^{k} \mathbf{v}_{i_k} \mathbf{v}_{i_k}^{H} (\mathbf{H}_{i_k}^{k})^{H} \left(\sum_{(l,j) \neq (k,i)} \mathbf{H}_{i_k}^{l} \mathbf{v}_{j_l} \mathbf{v}_{j_l}^{H} (\mathbf{H}_{i_k}^{l})^{H} + \sigma^2 \mathbf{I}_N \right)^{-1} \right|$$

$$\tag{3}$$

2.2   Power Consumption Model

The power consumption at a certain BS q_k can be modeled as

$$P_{q_k} = \xi P_{q_k}^{tx} + P_{q_k}^{sp,ct} + P_{q_k}^{bh} \tag{4}$$

where $P_{q_k}^{tx}$, $P_{q_k}^{sp,ct}$ and $P_{q_k}^{bh}$ denote the transmission power, signal processing power consumption and backhaul power consumption respectively, ξ is the reciprocal of power amplifier efficiency. Define $\boldsymbol{\Phi}_{q_k}$ as the row selection matrix which has all zeros except M ones on the main diagonal corresponding to the M antennas of BS q_k. Then, the total transmission power can be expressed by

$$P_{q_k}^{tx} = \sum_{i_k \in \mathcal{I}_k} \mathbf{v}_{i_k}^H \boldsymbol{\Phi}_{q_k}^H \boldsymbol{\Phi}_{q_k} \mathbf{v}_{i_k} \tag{5}$$

The capacity gain brought by MCT is accompanied with the increased power consumption. MCT introduces additional operation on each BSs, the signal to be transmitted should be exchanged by BSs through the backhaul and the joint signal processing is needed to suit joint transmissions. Hence, we refer to the power consumption model in [9] and the signal processing power is modeled to be a quadratic function of the cooperative cluster size as follows

$$P_{q_k}^{sp,ct} = p_{q_k}^{sp}(0.87 + 0.1\,|Q_k| + 0.03\,|Q_k|^2) \tag{6}$$

The backhaul power consumption is caused by data exchange in the cluster and is modeled as

$$P_{q_k}^{bh} = \frac{1}{C_{bh}}\left(\frac{2pq\,|Q_k|^2}{T_s}\right) \tag{7}$$

where C_{bh} denotes the backhaul capacity and T_s denotes the symbol period. p and q represent the additional pilot density and relevant signaling, respectively.

3 Energy Efficient Clustering and Beamforming

3.1 Problem Formulation

For kth cluster, the throughput can be written as

$$C_k(\{\mathbf{v}_{i_k}\}) = \sum_{i_k \in I_k} R_{i_k} \tag{8}$$

The total power consumption of k-th cluster is given by

$$P_k(\{\mathbf{v}_{i_k}\}) = \sum_{q_k \in Q_k} P_{q_k} = \xi \sum_{i_k \in I_k} \mathbf{v}_{i_k}^H \mathbf{v}_{i_k} + P_c \tag{9}$$

where P_c is the total circuit power consumption. Energy efficiency (EE) of the whole network is defined as $\mathrm{EE}(\{\mathbf{v}_{i_k}\}, Q) = \frac{\sum_{k=1}^K C_k(\{\mathbf{v}_{i_k}\})}{\sum_{k=1}^K P_k(\{\mathbf{v}_{i_k}\})}$. Hence, the energy efficient clustering and beamforming problem can be formulated as

$$\mathbf{P1}: \quad \max_{\{\mathbf{v}_{i_k}\}, Q} \quad \mathrm{EE}(\{\mathbf{v}_{i_k}\}, Q)$$

$$\text{s.t.} \quad \sum_{i_k \in I_k} \mathbf{v}_{i_k}^H \boldsymbol{\Phi}_{q_k}^H \boldsymbol{\Phi}_{q_k} \mathbf{v}_{i_k} \le p_{q_k}, \; \forall q_k \in \mathcal{Q} \tag{10}$$

The above problem is hard to solve, so we decouple the problem and use the hierarchical iterative algorithm to solve it. In outer iteration, the CFG is exploited to obtain the network partition Q. In inner iteration, the energy efficient beamforming problem is solved based on the given network partition Q.

3.2   Energy Efficient Beamforming

When the clusters are given, the origin problem can be rewritten as

$$\textbf{P2}: \quad \max_{\{\mathbf{v}_{i_k}\}} \quad EE(\{\mathbf{v}_{i_k}\}, Q)$$

$$\text{s.t.} \quad \sum_{i_k \in I_k} \mathbf{v}_{i_k}^H \boldsymbol{\Phi}_{q_k}^H \boldsymbol{\Phi}_{q_k} \mathbf{v}_{i_k} \le p_{q_k}, \ \forall q_k \in \mathcal{Q} \tag{11}$$

Proposition 1. *P2 has optimal objective value θ^* if and only if $f(\theta^*) = 0$, where univariate function $f : \mathbb{R} \longmapsto \mathbb{R}$ is defined as*

$$f(\theta) \triangleq \max_{\{v_{i_k}\}} \left\{ \sum_{k=1}^K C_k(\{\mathbf{v}_{i_k}\}) - \theta \sum_{k=1}^K P_k(\{\mathbf{v}_{i_k}\}) \right\}$$

$$\text{s.t.} \quad \sum_{i_k \in I_k} \mathbf{v}_{i_k}^H \boldsymbol{\Phi}_{q_k}^H \boldsymbol{\Phi}_{q_k} \mathbf{v}_{i_k} \le p_{q_k}, \ \forall q_k \in \mathcal{Q} \tag{12}$$

Proof. Based on the analysis in [7], we can conclude that $f(\theta)$ is a monotonically decreasing function of θ and the equation $f(\theta) = 0$ has a unique solution θ^*. Therefore if we find certain θ that makes the objective function of (12) equals to zero, then the corresponding beamformers are also the optimal beamformers of problem **P2**.

Note that the problem in (12) is hard to solve due to the non-convexity of capacity $C_k(\{\mathbf{v}_{i_k}\})$, we introduce a set of new weight variables $\{w_{i_k}\}$ for each user. Then the problem can be reformulated as:

$$\textbf{P3}: \quad \max_{\{\mathbf{v}_{i_k}\}, \{\mathbf{u}_{i_k}\}, \{w_{i_k}\}} \quad \sum_k \sum_{i_k \in \mathcal{I}_k} (\log(w_{i_k}) - w_{i_k} e_{i_k}) - \theta \sum_k \left(\xi \sum_{i_k \in \mathcal{I}_k} \mathbf{v}_{i_k}^H \mathbf{v}_{i_k} + P_c \right)$$

$$\text{s.t.} \quad \sum_{i_k \in \mathcal{I}_k} \mathbf{v}_{i_k}^H \boldsymbol{\Phi}_{q_k}^H \boldsymbol{\Phi}_{q_k} \mathbf{v}_{i_k} \le p_{q_k}, \ \forall q_k \in \mathcal{Q}$$

$$e_{i_k} \text{ is given by (2).}$$

$$\tag{13}$$

Similar to [6,10], we can conclude that if $(\{\mathbf{v}_{i_k}^*\}, \{\mathbf{u}_{i_k}^*\}, \{w_{i_k}^*\})$ is the optimal solution to **P2**, then $\{\mathbf{v}_{i_k}^*\}$ must be the optimal solution to **P1** and (12). Conversely if $\{\mathbf{v}_{i_k}^*\}$ is the optimal solution of **P2** and (12), then $(\{\mathbf{v}_{i_k}^*\}, \{\mathbf{u}_{i_k}^*\}, \{w_{i_k}^*\})$ must be the optimal solution to **P2**, where

$$\mathbf{u}_{i_k}^* = \boldsymbol{\Sigma}_{i_k}^{-1}(\{\mathbf{v}_{i_k}^*\}) \mathbf{H}_{i_k}^k \mathbf{v}_{i_k}^*$$

$$w_{i_k}^* = \left(1 - (\mathbf{v}_{i_k}^*)^H (\mathbf{H}_{i_k}^k)^H \boldsymbol{\Sigma}_{i_k}^{-1}(\{\mathbf{v}_{i_k}^*\}) \mathbf{H}_{i_k}^k \mathbf{v}_{i_k}^*\right)^{-1} \tag{14}$$

with $\boldsymbol{\Sigma}_{i_k}(\{\mathbf{v}_{i_k}^*\}) = \sum_{(l,j)} \mathbf{H}_{i_k}^l \mathbf{v}_{j_l} \mathbf{v}_{j_l}^H (\mathbf{H}_{i_k}^l)^H + \sigma^2 \mathbf{I}$.

In what follows, we solve **P3** for given θ, $\{\mathbf{u}_{i_k}\}$ and $\{w_{i_k}\}$, which is a convex optimization problem. the Lagrangian function of **P3** can be stated as

$$\mathcal{L}(\{\mathbf{v}_{i_k}\}, \{\lambda_{q_k}\}) = \sum_{k=1}^{K} \sum_{i_k \in I_k} w_{i_k} \left(1 + \sum_{(l,j)} \mathbf{u}_{i_k}^H \mathbf{H}_{i_k}^l \mathbf{v}_{j_l} \mathbf{v}_{j_l}^H (\mathbf{H}_{i_k}^l)^H \mathbf{u}_{i_k} - \mathbf{u}_{i_k}^H \mathbf{H}_{i_k}^k \mathbf{v}_{i_k} - \mathbf{v}_{i_k}^H (\mathbf{H}_{i_k}^k)^H \mathbf{u}_{i_k}\right)$$
$$+ \theta\xi \sum_{k=1}^{K} \sum_{i_k \in I_k} \mathbf{v}_{i_k}^H \mathbf{v}_{i_k} + \sum_{q_k \in \mathcal{Q}} \lambda_{q_k} \left(\sum_{i_k \in I_k} \mathbf{v}_{i_k}^H \boldsymbol{\Phi}_{q_k} \mathbf{v}_{i_k} - p_{q_k}\right)$$

$$(15)$$

where $\{\lambda_{q_k}\}$ denote the Lagrange multipliers associated with the power constraints. Applying Lagrangian dual theory and the Karush-Kuhn-Tucker (KKT) conditions, the optimal beamformer is given by

$$\mathbf{v}_{i_k}^* = w_{i_k} \left(\sum_{(l,j)} (\mathbf{H}_{j_l}^k)^H \mathbf{u}_{j_l} \mathbf{u}_{j_l}^H \mathbf{H}_{j_l}^k + \theta\xi\mathbf{I} + \sum_{q_k \in \mathcal{Q}_k} \lambda_{q_k} \boldsymbol{\Phi}_{q_k}^H \boldsymbol{\Phi}_{q_k}\right)^{-1} (\mathbf{H}_{i_k}^k)^H \mathbf{u}_{i_k}$$

$$(16)$$

The optimal Lagrange multipliers $\{\lambda_{q_k}^*\}$ can be obtained by gradient method. Then we develop Algorithm 1 as below to numerically search the optimal value of θ and $\{\mathbf{v}_{i_k}\}$.

Algorithm 1. Energy efficient beamforming

1: Initialize θ^0
2: Initialize $\{\mathbf{v}_{i_k}^0\}$
3: **repeat**
4:     Sequentially update $\mathbf{u}_{i_k}^t, w_{i_k}^t$
5:     update $\mathbf{v}_{i_k}^{t+1}$ and θ^{t+1}
6: **until** certain stopping criteria met.

3.3   Energy Efficient Clustering as a Coalition Formation Game

We define the energy efficient cluster forming game (EECFG) as a triplet, $\mathbb{G}_{\text{EECF}} = (\mathcal{Q}, u, Q)$ in a characteristic form. The players, namely BSs, are affected each other through mutual interference, and they seek to form cooperative clusters to improve energy efficiency. Moreover, u is a characteristic function that quantifies the value of a coalition. $Q = (Q_1, \ldots, Q_k, \ldots, Q_K)$ which satisfies $\forall k_1, k_2 \in \{1, \ldots, K\}, Q_{k_1} \cap Q_{k_2} = \varnothing, \bigcup_{k=1}^{K} Q_k = \mathcal{Q}$ is a partition of $\mathcal{Q}$ and shows a cooperative structure of the network. Coalition value set is defined as

$$u(Q_k) = \{\boldsymbol{v}(Q_k) \in \mathbb{R}^{|Q_k|} | v_b(Q_k), \forall b \in Q_k\} \qquad (17)$$

where $v_{q_k}(Q_k)$ is an element of $\boldsymbol{v}(Q_k)$ and represents the utility that player $q_k \in Q_k$ can obtain in the coalition Q_k. Here, we define the BS' utility as the EE it achieves when serving attached users. Note that the utility obtained by each BS in Q_k depends on the joint strategies that all BSs in Q_k select and the coalition value cannot be arbitrarily apportioned among the members. Hence, the proposed $\mathbb{G}_{\mathrm{EECF}}$ has a nontransferable utility (NTU). Therefore, we adopt merge and split rules to obtain the optimal solution of $\mathbb{G}_{\mathrm{EECF}}$. The modified merge and split rules are defined as follows

Definition 1. *Merge rule: merge any two coalitions* Q_{k_1}, Q_{k_2}, *if* $\{Q_{k_1} \cup Q_{k_2}\} \rhd_p$
$\{Q_{k_1}, Q_{k_2}\} \begin{cases} \{Q_{k_1} \cup Q_{k_2}\} \rhd_p \{Q_{k_1}, Q_{k_2}\} \\ \{Q_{k_1} \cup Q_{k_2}\} \notin h(\bar{Q}) \end{cases}$;
Split rule: split any coalitions $\{Q_{k_1}, Q_{k_2}\}$ *into two coalitions* Q_{k_1}, Q_{k_2}, *if* $\{Q_{k_1} \cup$
$Q_{k_2}\} \rhd_p \{Q_{k_1}, Q_{k_2}\} \begin{cases} \{Q_{k_1}, Q_{k_2}\} \rhd_p \{Q_{k_1} \cup Q_{k_2}\} \\ \{Q_{k_1}, Q_{k_2}\} \notin h(\bar{Q}) \end{cases}$, *where* $\rhd_p$ *denotes the pareto order,* $h(\bar{Q})$ *denotes the history clustering information.*

The network partition is first initialized to $Q^0 = \{\{1\}, \ldots, \{|\mathcal{Q}|\}\}$, i.e., each BS separately serves its attached users. When the ICI is severe, the BSs have the incentive to cooperate with dominating interferer to jointly serve users, thus the EE is improved. After initialization, merge and split operations are performed to iteratively obtain the optimal partitions. At each iteration, the proposed energy efficient beamforming algorithm (Algorithm 1) is used to determine the cooperative transmission strategy and calculate the achieved EE. In this way, we can obtain the EE optimal clusters and relevant beamformer through hierarchical iteration.

Since the total number of the partitions is finite, i.e., Bell number and history clustering information is introduced into the algorithm to avoid the repetitive deviations, the proposed cluster forming algorithm always converges with any initial partition. In addition, according to Theorem 6.2 [2], the proposed cluster forming algorithm is $\mathbb{D}_{hp}$-stable.

4   Numerical Results

In this section, performance of the proposed energy efficient clustering and beamforming algorithm for cooperative multicell networks are numerically evaluated. As shown in Fig. 1(a), we consider a multicell systems with 7 uniformly distributed BSs each equipped with 2 antennas. Users are attached to BSs whose pilot signal is strongest. At each time slot, we assume that each BS only serves one single-antenna user. The channel from BS q_l to user i_k is assumed to be $\mathbf{H}_{i_k}^{q_l} = \sqrt{f_{i_k,ls}^{q_l}} \mathbf{H}_{i_k,ss}^{q_l}$, where $\mathbf{H}_{i_k,ss}^{q_l}$ is small scale fading coefficient and is modeled as the Gaussian distribution with zero mean and unit covariance. $f_{i_k,ls}^{q_l}$ represents the large scale fading coefficient and is modeled as $f_{i_k,s}^{q_l} - 15.3 - 37.6 \log_{10}(d_{i_k}^{q_l})$, where $f_{i_k,s}^{q_l}$ denotes the shadow fading in decibels. For comparison, some baseline strategies are also simulated: (1) the proposed energy efficient beamforming algorithm with no cooperation. (2) the proposed energy efficient beamforming algorithm with full cooperation. (3) the maximum ratio transmission (MRT) with no cooperation. (4) MRT with proposed clustering algorithm.

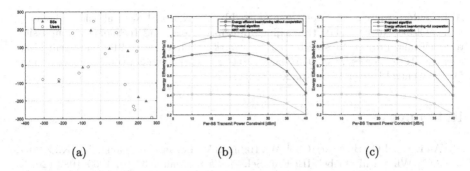

(a)   (b)   (c)

Fig. 1. (a) Simulated cooperative multicell network model (a) Energy efficiency vs transmit power constraint: when flexible cooperation is considered (b) Energy efficiency vs transmit power constraint: comparison with full cooperation case.

Figure 1(b) shows the EE with different transmit power constraint. It can be seen that when applying the proposed energy efficient beamforming algorithm, the EE can be greatly improved compared with MRT case and the gain can be up to 104%. Moreover, when the CFG based clustering algorithm is also exploited, the EE can be further improved by 20.5%. In addition, we can see that with the increase of the transmit power constraint, the EE first increases and then decreases. This is due to the increased ICI and transmit power. Figure 1(c) shows the EE performance with different degree of cooperation. Full cooperation of multicells will improve the network capacity, however, it leads to increased energy consumption, and hence results in EE degradation. Comparing with full cooperation, the proposed CFG based clustering algorithm enables more flexible cooperation. When the ICI is severe, BSs will cooperate with aggressor BSs to improve the EE. Hence the proposed algorithm shows superior performance against the full cooperation.

5   Conclusion

In this paper, we considered joint clustering and beamforming for energy efficiency optimization in cooperative multicell networks. In the dense network scenario, each BSs separately serve their attached UEs may not be energy efficient due to the severe intercell interference. So some BSs may be prompted to cooperative with interfering neighbor BSs in order to improve the energy efficiency. Hence in this paper, the hierarchical iterative strategy is proposed to fulfill the goal of flexible cooperation. The EE optimization problem is divided into two coupled subproblem. CFG is used to obtain the EE-optimal network partition. For beamforming problem, based on the fractional program and the MMSE model, the EE optimization problem is transformed to a convex optimization problem and is efficiently solved. For the future work, we will consider the user scheduling and extend the algorithm into heterogeneous cloud radio access networks.

Acknowledgment. This work is supported by the National Natural Science Foundation of China, No. 61271179, the Beijing Municipal Science and Technology Commission research fund project "Research on 5G Network Architecture and Its Intelligent Management Technologies", No. D151100000115002, and the Fundamental Research Funds for the Central Universities, No. 2014ZD03-01.

References

1. Andrews, J.G., Buzzi, S., Choi, W., Hanly, S.V., Lozano, A., Soong, A.C.K., Zhang, J.C.: What will 5G be? IEEE J. Sel. Areas Commun. **32**(6), 1065–1082 (2014). doi:10.1109/JSAC.2014.2328098
2. Apt, K.R., Radzik, T.: Stable partitions in coalitional games. arXiv preprint cs/0605132 (2006)
3. Haijun, Z., Hui, L., Chunxiao, J., Xiaoli, C., Nallanathan, A., Xiangming, W.: A practical semidynamic clustering scheme using affinity propagation in cooperative picocells. IEEE Trans. Veh. Technol. **64**(9), 4372–4377 (2015). doi:10.1109/TVT.2014.2361931
4. He, S., Huang, Y., Yang, L., Ottersten, B., Hong, W.: Energy efficient coordinated beamforming for multicell system: duality-based algorithm design and massive mimo transition. IEEE Trans. Commun. **63**(12), 4920–4935 (2015). doi:10.1109/TCOMM.2015.2496948
5. Heliot, F., Imran, M.A., Tafazolli, R.: Energy efficiency analysis of idealized coordinated multi-point communication system. In: 2011 IEEE 73rd Vehicular Technology Conference (VTC Spring), pp. 1–5 (2011). doi:10.1109/VETECS.2011.5956410
6. Hong, M., Sun, R., Baligh, H., Luo, Z.Q.: Joint base station clustering and beamformer design for partial coordinated transmission in heterogeneous networks. IEEE J. Sel. Areas Commun. **31**(2), 226–240 (2013). doi:10.1109/JSAC.2013.130211
7. Jagannathan, R.: On some properties of programming problems in parametric form pertaining to fractional programming. Manage. Sci. **12**(7), 609–615 (1966)
8. Lozano, A., Heath, R.W., Andrews, J.G.: Fundamental limits of cooperation. IEEE Trans. Inf. Theory **59**(9), 5213–5226 (2013)
9. Pantisano, F., Bennis, M., Saad, W., Verdone, R., Latva-aho, M.: On the dynamic formation of cooperative multipoint transmissions in small cell networks. In: 2012 IEEE Globecom Workshops (GC Wkshps), pp. 1139–1144 (2012). doi:10.1109/GLOCOMW.2012.6477739
10. Qingjiang, S., Razaviyayn, M., Zhi-Quan, L., Chen, H.: An iteratively weighted mmse approach to distributed sum-utility maximization for a mimo interfering broadcast channel. IEEE Trans. Signal Process. **59**(9), 4331–4340 (2011). doi:10.1109/TSP.2011.2147784
11. Saad, W., Han, Z., Debbah, M., Hjørungnes, A., Başar, T.: Coalitional game theory for communication networks. IEEE Signal Process. Magaz. **26**(5), 77–97 (2009)
12. Somekh, O., Simeone, O., Bar-Ness, Y., Haimovich, A.M., Shamai, S.: Cooperative multicell zero-forcing beamforming in cellular downlink channels. IEEE Trans. Inf. Theory **55**(7), 3206–3219 (2009). doi:10.1109/TIT.2009.2021371
13. Soret, B., Hua, W., Pedersen, K.I., Rosa, C.: Multicell cooperation for LTE-advanced heterogeneous network scenarios. IEEE Wirel. Commun. **20**(1), 27–34 (2013). doi:10.1109/MWC.2013.6472196

14. Wei, X., Yuke, C., Hua, Z., Li, G.Y., Xiaohu, Y.: Robust beamforming with partial channel state information for energy efficient networks. IEEE J. Sel. Areas Commun. **33**(12), 2920–2935 (2015). doi:10.1109/JSAC.2015.2478720
15. Zengfeng, Z., Lingyang, S., Zhu, H., Saad, W.: Coalitional games with overlapping coalitions for interference management in small cell networks. IEEE Trans. Wirel. Commun. **13**(5), 2659–2669 (2014). doi:10.1109/TWC.2014.032514.130942
16. Zhang, H., Chu, X., Guo, W., Wang, S.: Coexistence of wi-fi and heterogeneous small cell networks sharing unlicensed spectrum. IEEE Commun. Magaz. **53**(3), 158–164 (2015). doi:10.1109/MCOM.2015.7060498
17. Zhang, H., Jiang, C., Cheng, J., Leung, V.C.M.: Cooperative interference mitigation and handover management for heterogeneous cloud small cell networks. IEEE Wirel. Commun. **22**(3), 92–99 (2015). doi:10.1109/MWC.2015.7143331
18. Zhang, H., Jiang, C., Mao, X., Chen, H.H.: Interference-limited resource optimization in cognitive femtocells with fairness and imperfect spectrum sensing. IEEE Trans. Veh. Technol. **65**(3), 1761–1771 (2016). doi:10.1109/TVT.2015.2405538

Cross-Monotonic Game for Self-organized Context-Aware Placement of Services with Information Producers and Consumers

Manuel Osdoba[(✉)] and Andreas Mitschele-Thiel

Integrated Communication Systems Group,
Technische Universität Ilmenau, Ilmenau, Germany
{manuel.osdoba,andreas.mitschele-thiel}@tu-ilmenau.de
http://www.tu-ilmenau.de/iks

Abstract. Deploying service instances in a network requires multiple considerations. Firstly, an instance should be placed near clients that need the service. Secondly, it should be in the centre of those clients. Thirdly, the service provider himself should benefit from placing additional service instances.

We approximate problem one and two by a distributed auction. The winners of the auction are agents that bid to join in a cost-sharing scheme with cross-monotonic cost shares to solve problem three. Service instances that have an appropriate number of clients that consume may serve a context. Those with clients that are passive or mainly produce information may not serve the context because they can not pay their cost shares and thus would not be beneficial to the service provider. Clients of those instances (producers) are directly connected to the service providers central server. Our algorithm fits well to services with a regular consumer/producer ratio of 0.75/0.25.

Keywords: Service placement · Self-organization · Sparse knowledge · Multicast game · Facility location problem · K-mean/median-problem

1 Introduction

Resources in mobile networks are limited and the concurrent transfer of duplicate data is a waste of resources. If the flood of information is big enough, every network will suffer from the concurrent transfer of duplicate data. Therefore, it is of use to replicate or cache data at distinct points in the network. However, replicas introduce additional synchronization traffic to keep the different copies consistent. As long as the content is not changed, no synchronization mechanism is triggered. We call nodes that *read* data *consumers*. On the other hand, there are nodes that *alter/write* data. Those are *producers*. Updated information is assumed to be propagated through the entire network until all active instances that store the information received it.

Nowadays algorithms replace static mechanisms and human intervention in networks. A cloud infrastructure is usually controlled by central administrative

© ICST Institute for Computer Sciences, Social Informatics and Telecommunications Engineering 2017
J. Cheng et al. (Eds.): GameNets 2016, LNICST 174, pp. 34–42, 2017.
DOI: 10.1007/978-3-319-47509-7_4

entities that orchestrate an entire network. However, clouds as well as service- and content delivery are also of relevance in wireless networks and in those net- works, entities have sparse knowledge on the topology. That was our motivation do develop a self-organized algorithm that deals with the challenges that were stated in the abstract.

The paper is organized as follows. In Sect. 2, we study previous approaches that influenced our mechanism. In the following part, our model and assump- tions as as well as definitions on cost sharing games and requirements are given. Section 3 deals with our cross-monotonic algorithm, which is evaluated in Sect. 5. We conclude the results, shortcomings and advantages in Sect. 6.

2   Related Work

Our approach addresses two challenges in future networks. Service instances need to be placed near clients that request the service. Usually, a service provider analyzes the clients behavior and the decision where to place service instances is based upon an offline optimization. There were several approaches that addressed an online optimization of wireless (sensor) networks, like [11].

Games for Service Placement. In wireless networks, environmental factors need to be monitored. For that purpose, Wu et al. [11] created a submodular game that improves Quality of Monitoring. Once a wireless sensor node notices a lack of monitoring, it solves a knapsack problem to find out, whether it has enough resources to run an additional monitoring application. An appropriate per-node utility function considers a neighbors allocation and maximizes the social welfare that is measured by the quality of monitoring. Whenever a node changes its strategy, it has to send its modified strategy to its neighbors.

Different kinds of facility location and k-Median problems were addressed by Pál et al. in [8]. In their Single-Source Rent-or-Buy game, a strategy-proof cost-sharing scheme was proposed. The edges of a Steiner tree are bought while edges on the shortest path from a receiver to the Steiner tree are rented. Every individual packet transfer along a rented edge has to be paid separately. The costs that arise if an edge is bought, are *shared* among the receivers in the Steiner component. Furthermore, their ghost-mechanism requires the receivers to continuously fund a fraction of their cost shares to establish the Steiner tree.

The second challenge is to describe and identify the context of nodes.

Context-Awareness. Self-Optimization and Self-Configuration is an essential functionality in the Internet of Things. Several *things* can share a context. In [9] is described how the correlation among contexts can be measured. Jaffal et al. [3] analyzed, in which way a context can be abstractly described to aid in the design of pervasive systems. Najar et al. [5] elaborated on how a service, that satisfies a clients intention in a given context, can be chosen. Especially [1,6] analyze an architecture that includes contextual awareness as a factor in deploying and designing services.

3   System Model and Requirements

Data for *consumers* can be retrieved from the node that stores all data or at lower cost (and lower delay) from a nearby service instance. If the data item was previously retrieved, *consumers* get their information from a nearby service instance and the retrieval from the central node can be omitted. A high cache hit rate results in a decreased use of the connection to the central node. It increases if it is likely that clients request same data and that is the case if those are correlated. We see two ways to find out whether data is correlated.

Firstly, data can be partitioned. Hence, requests are correlated if clients request data from the same partition. Secondly, for unpartitioned/unstructured data, we use methods of big data and data analysis. The information, a client is interested in can be put into a selector vector. It has to be staged by the service/content provider (e.g. by formal concept analysis [3]) and is used by the facilities to discriminate clients within the same context from those who are not. The comparison of a selector vector with a client vector can be done with the symmetric Kullback-Leibler- or symmetric Jensen-Shannon divergence [1–3,5,6,9]. Matching a clients behavior to the selector vector may require deep packet inspection on samples of the traffic. Furthermore, a candidate needs to be aware of the clients profile in the service. These can be *information consumers* or *information providers*. The occurring system costs are shown in Eq. 1

$$\Omega(S \subseteq F, t) = \sum_{v \in S} \sum_{c \in C_v} \left(r_c(t) \cdot (\mathrm{cost}(\mathrm{P}_{c,s}) - \mathrm{cost}(\mathrm{P}_{c,v})) + w_c(t) \mathcal{C}_{Steiner} \right) \quad (1)$$

In Fig. 1, all clients have to retrieve their data from the central instance s. If node u in Fig. 1 *reads* data, it incurs cost of $\sum_{e \in P_{us}} \mathrm{cost}(e) = 18$. If he acts as an producer and *writes* to s, costs will be 18, equally. If the same scenario occurs in Fig. 2, node u updates data that is distributed among the nodes that are colored green, it incurs cost of 43. If another user requests the updated content, the data is retrieved from the local instance. If $\lceil (40 - 15)/20 \rceil = 2$ users in the

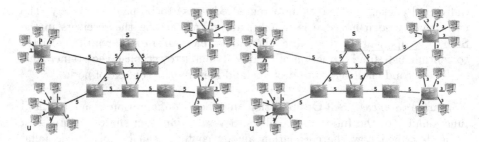

Fig. 1. Only one service instance is active (=green): Configuration if majority of operations is *Write* (Color figure online)

Fig. 2. Several service instances are activated (=green) and connected by Multicast Tree: configuration if majority of operations is *Read* (Color figure online)

lower right corner of the picture read the new content, costs are compensated. Our approach suits networks in which nodes have sparse knowledge. Therefore, centralized algorithms are inappropriate. To find nodes that are appropriate to serve clients, the k-Median problem is approximated with best response dynamics. If consumers and producers change their behavior, service instances adapt accordingly. Consumers at a service instance compensate the costs that arise on the reception of the updates. If the amount of consumers is too low, those costs are not compensated and the instance stops serving that context. We optimize both with a cost sharing scheme.

Definition 1 (Cost-Sharing Scheme [7]). *A cost-sharing scheme is a function* $\xi : A \times 2^A \rightarrow \mathbb{R}^+ \cup \{0\}$ *such that, for every* $S \subset A$ *and every* $i \notin S, \xi(i, S) = 0$.

The value $\xi(i, S)$ determines the cost shares of the agent i within the set S. Usually an agent has an incentive to cover the costs it invests in cost shares. An agents revenue/gain has to be greater than its investment in the cost shares. An agent that can not cover its cost shares is pruned from the scheme, which is one important mechanism to achieve the following property of our game.

Definition 2 (Cross-Monotonicity [7]). *A cost-sharing scheme* ξ *is cross-monotonic if for all* $S, T \subseteq A$ *and* $i \in S, \xi(i, S) \geq \xi(i, S \cup T)$.

Cross-monotonicity is also known under the term population monotonic [10]. The set of participating agents increases while the cost shares decrease. An agents cost share *may not rise* even if additional agents join in. An equivalent argument for cross-monotonicity is $\xi(i, S) \geq \xi(i, S')$ for all $S \subseteq S'$. Therefore cross-monotonicity stimulates other agents to join. Cost sharing schemes should provide further import characteristics like *competitiveness* and *cost recovery*.

Definition 3 (Competitiveness [8]). $\sum_{i \in S} \xi(S, i) \leq c^*(S)$ *and assures that the participating agents are not charged more than the true cost* $c^*(S)$.

If competitiveness is not assured, there would be the possibility, that some other agent could offer the service at lower cost. The following term is also known as weak budget-balance.

Definition 4 (Cost Recovery [8]). $\sum_{i \in S} \xi(i, S) \geq c^*(S)$ *and assures that the costs are recovered.*

Our game is cross-monotonic and recovers the cost of the update distribution.

4   Cross-Monotonic Semi-cost Recovering Game

A game is a triple of (A, S, u_v) with agents A, a strategyspace S and a revenue function u_v for node v. As mentioned in the previous section, the agents are the available facilities that can host an instance of the service. The strategy, an agent may choose from, is defined as follows:

$$\mathcal{S} = \{2^C \times \mathbb{R} \times \mathbb{B}\}$$

An agent vs strategy for k correlated sets is $\mathfrak{s} = (C' \subseteq C_v, \mathfrak{b}, \mathfrak{a})^k \in \mathcal{S}^k$. In $\mathfrak{s}$, C' represents the correlated clients of a context that are near node v, $\mathfrak{b}$ is the nodes bid to serve the client set C' and $\mathfrak{a}$ is true if the clients $read/write$ ratio fulfills Eq. 2 and therefore, the instance is efficient in decreasing the value of Eq. 1. In Eq. 2, value $\alpha_v(S, i, t)$ denotes the agents cost-shares that have to be paid at every reception of an update in context i and is depicted in Eq. 4. Agent vs strategy for k correlated sets is $\mathfrak{s} = ((C'_0 \subseteq C_v, \mathfrak{b}_0, \mathfrak{a}_0), \ldots, (C'_{k-1} \subseteq C_v, \mathfrak{b}_{k-1}, \mathfrak{a}_{k-1})) \in \mathcal{S}^k$. The available contexts inside the network are denoted by $\mathcal{K}$.

$$\sum_{c \in C'} \int_t^{t+\tau} r_c(t) \cdot (\text{cost}(P_{c,s}) - \text{cost}(P_{c,v})dt > \sum_{c \in C' \setminus C_v} \int_t^{t+\tau} w_c(t) \cdot \alpha_v(S)dt \quad (2)$$

Algorithm 1 runs on every agent $v \in \mathcal{A}$. Per correlated set of clients (line 2), a node determines its median distance to the clients (line 3). The nodes bid value $\mathfrak{b}$ to serve the clients is computed with a Gaussian. We choose $\mu = 0$ and $\sigma = |C_v|$ (line 3). The argument of the Gaussian is calculated as depicted in line 4. Several approaches to the Facility Location Problem (e.g. [8]) construct a ball around the facilities or clients. At the intersection point of several balls, a facility is opened. In this approach, the facilities are agents.

A node estimates its maximum cost shares within a context with Eq. 3. $\mathfrak{C} : \mathcal{K} \to C$ represents the clients within the context.

$$\mathfrak{B}_v(i, t) = \sum_{c \in (\mathfrak{C}(i) \cap C_v)} r_c(t) \cdot \text{cost}(P_{c,s} - P_{c,v}) \quad (3)$$

Node v sends its bid $\mathfrak{B}_v(i, t)$ to serve context i to the central instance. At this point, we use scheme [7, Sect. 14.2.2, p. 367]. The central service instance decides whether the offered bid covers the cost to integrate node v into the Multicast tree. It is handled that way because we expect the central instance to know about the current $read/write$ ratio in context i. Node vs cost shares that were determined by the referenced scheme are depicted in Eq. 4. Here, $\text{ST}(i), i \in \mathcal{K}$ is the current multicast tree of the service instances that serve context i.

$$\alpha_v(S, i, t) = \min_{w \in ST(i)} \left\{ \text{cost}(P_{v,s}), \text{cost}(P_{v,w}) + \text{cost}^{ST(i)} \left(P_{w,s}^{ST(i)} \right) \right\} \quad (4)$$

Every service instance is charged the cost shares $\alpha_v(S, i, t)$ (Eq. 4) per received update. The path of node w to s in the Multicast tree is denoted as $P_{w,s}^{ST}$ and the proportionate cost of node w as cost^{ST}. The cost shares $\alpha_v(S, i, t)$ are cross-monotonic (Definition 2) if the triangle inequality holds. Furthermore, the mechanism in line 10 ascertains that the recovery of the cost of propagating the updates (Definition 4) can not be violated longer than for time τ.

```
 1  if Received strategy s' or topology change or consumer/producer change then
 2      for C' ∈ correlated_sets(v) do
 3          m ← median d(c, v);  σ ← |C'|;
                   c∈C'
 4          δ ← ∑ (m − d(c, v));
               c∈C'
                       −δ²
 5          b ← √|C'| · e^(2σ²) − ρ + r;
 6          s ← (C', b, False);
 7          if b > 0 or b > b_s' then
 8              if strategy s changes then
 9                  Send s to neighbors; sleep(θ);
10                  if Eq. 2 holds then
11                      Connect all c ∈ C' to v
12                  else
13                      ∅-Strategy
14                  end
15              end
16          else
17              ∅-Strategy
18          end
19      end
20  end
```

Algorithm 1. Best response mechanism for node v

5 Evaluation

For evaluation, we investigated different *consumer/producer* profiles.

Simulation Setup. The simulations were implemented in Python with SciPy, NumPy and SimPy. For the computation of the Steiner tree among the active facilities, we used the submodular function optimizer library [4]. User operations follow a Poisson process. Each of the 80 *clients* executes a *read* or *write* operation with distinct probabilities. Exact pairs of *read/write* ratios are shown on the x-axis in Fig. 3. All users are homogeneous, meaning all users have a common *read/write*-ratio and Poisson arrival rate of 2. A node may change an item if it has previously read it. The performance is stable regarding the number of requested items or number of operations per time slot.

Results. Figure 3 shows that in the range of *read/write* ratios between 0.05/0.95 and 0.35/0.65, the cost of the single-server (=blue) and multi-server configuration (=red) rise fast. An item has to be read before it can be changed. Therefore, *write* cost follow *read* and are upper bounded by the *read* cost in the single-server case. That is not the case in the multi-server solution. If an item is already available at the service instance a client is connected to, the item is retrieved from the local instance. An item is not available at a local service instance if no client

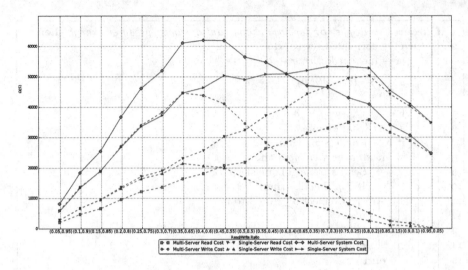

Fig. 3. Cost for different read/write profiles (Color figure online)

previously read it *or* no other client in the same context inside the *entire* network changed it. If another client would have changed it, the update was propagated. In our simulation, the multi-server cost are clearly above the single-server approach in the range of *read/write* ratios between 0.35/0.65 to 0.5/0.5. From 0.5/0.5, the multi-server system cost decrease. At the ratio 0.75/0.25, our self-organized system has its optimal working point. Prior to the working point, line 10 in Algorithm 1 assures cost slightly above the single-server approach. Until ratio 0.6/0.4, Algorithm 1 enforces the single-server solution. Winners of the distributed auction to serve the context can not afford the cost shares. Therefore, clients request and write data directly to the single server. After that break-even point, service instances start from being able to pay their cost shares to receive updates and our mechanism causes the convergence to the multi-server configuration. At the working point, the service can be delivered at the same cost a single-server approach would have. However, the users have a lower latency. After the break-even point, multi-server configuration outruns the single-server approach.

Though, at the ratio 0.6/0.4, our algorithm violates the cost-recovery property for time τ. That is a disadvantage for service instances with a low number of clients. Those easily run in the situation where they can not cover their cost shares anymore. The mechanism in Algorithm 1, line 10 shuts down those service instances. Their revenue function stabilized at *read/write* ratios of 0.75/0.25. Therefore, our self-organized algorithm fits well to services, that regularly have 75% *consumers* and 25% *producers*.

6   Conclusion

This paper presents a cross-monotonic cost sharing scheme that solves the Service Placement Problem with *information producers* and *information consumers*. It is applicable for wireless as well as for IoT infrastructures. Our approach transforms a global cost function (Eq. 1) into a local optimization problem (Eq. 2) that is optimized by a self-organized algorithm. Thus, our algorithm runs and decides in each node solely based on local knowledge. Our approach shows a favorable working point at a *read/write* ratio of 0.75/0.25. The service can be delivered at the same cost a single-server approach achieves while the transfer of the up-to-date data is achieved at a considerably lower delay.

Self-organized systems benefit from cross-monotonic cost-sharing schemes. Cross-monotonicity assures that newly entering agents can not increase the cost shares of other nodes. That makes it ideal for service providers because a joining agent can not increase the cost shares of other agents. Our mechanism automatically shuts down entities that suffer from a lack of *consumers* and therefore provides an efficient mechanism in dealing with an unbalance between *information consumers* and *information providers*. It is a derivate of the cross-monotonic Multicast Game. However, an agents revenue changes over time and if the revenue becomes negative, our mechanism shuts down the agent. The mechanism allows agents to violate the cost-recovery property for a time τ.

In our future work, we will give proof on the competitiveness and α-cost-recovery in dependence of τ and real-world client-profiles. Furthermore, we will compare our scheme to connected facility location algorithms that perform an offline optimization. Additionally, the bidding mechanism for the approximation of the K-Median problem has to be compared to Pál et al. [8] Primal/Dual mechanism.

References

1. Bauer, C., Dey, A.K.: Considering context in the design of intelligent systems: current practices and suggestions for improvement. J. Syst. Softw. **112**, 26–47 (2016)
2. Jaffal, A., Kirsch-Pinheiro, M., Le Grand, B.: Unified and conceptual context analysis in ubiquitous environments. In: International Academy, Research, and Industry Association (IARIA), vol. 8, pp. 48–55 (2014)
3. Jaffal, A., Le Grand, B., Kirsch-Pinheiro, M.: Refinement strategies for correlating context and user behavior in pervasive information systems. Procedia Comput. Sci. **52**, 1040–1046 (2015)
4. Krause, A.: Submodular function optimization (2008). https://las.inf.ethz.ch/sfo/
5. Najar, S., Pinheiro, M.K., Souveyet, C.: Service discovery and prediction on pervasive information system. J. Ambient Intell. Hum. Comput. **6**(4), 407–423 (2015)
6. Naqvi, S.N.Z., Ramakrishnan, A., Preuveneers, D., Berbers, Y.: Walking in the clouds: deployment and performance trade-offs of smart mobile applications for intelligent environments. In: International Conference on Intelligent Environments (IE13), 16–19 July 2013, Athens, Greece (2013)

7. Nisan, N., Roughgarden, T., Tardos, E., Vazirani, V.V.: Algorithmic Game Theory. Cambridge University Press, New York (2007)
8. Pál, M., Tardos, É.: Group strategy proof mechanisms via primal-dual algorithms. In: 44th Symposium on Foundations of Computer Science (FOCS 2003), Proceedings, 11–14 October 2003, Cambridge, MA, USA, pp. 584–593 (2003)
9. Ramakrishnan, A., Preuveneers, D., Berbers, Y.: Enabling self-learning in dynamic and open IOT environments. Procedia Comput. Sci. **32**, 207–214 (2014)
10. Tazari, S.: Cross-monotonic cost-sharing schemes for combinatorial optimization games: a survey: course Project, CPSC 532A Multiagent Systems. University of British Columbia, Vancouver, Canada (2005)
11. Wu, C., Xu, Y., Chen, Y., Lu, C.: Submodular game for distributed application allocation in shared sensor networks. In: Proceedings of the IEEE INFOCOM 2012, 25–30 March 2012, Orlando, FL, USA, pp. 127–135 (2012)

Game Models and Theories

Tracking Areas Planning with Cooperative Game in Heterogeneous and Small Cell Networks

Lei Ning, Zhenyong Wang$^{(\boxtimes)}$, and Qing Guo

Communication Research Center, Harbin Institute of Technology, Harbin, China
{lning,zywang,qguo}@hit.edu.cn

Abstract. Increasing demands of data transmissions are promoting the acceleration of peaking rate per terminal especially in hot-spots. Numerous irregular deployments of small cells require efficient TA planning method in heterogeneous cellular networks. Macrocells preferred access is not a fundamental solution for TA planning, result from reducing the offloading ability of small cells. In this paper, a novel TA planning algorithm based on cooperative games is proposed by detecting similar communities. Simulation results show that it can reduce the signalling overhead while maintaining the utilization proportion of femtocells.

Keywords: Heterogeneous and small cell networks · Location management · Tracking areas planning · Cooperative game

1 Introduction

In recent years, various mobile terminals with high performance have captured the market rapidly all over the world. This leads a new era that varieties of services such as cloud computing, multimedia broadcasting, social networks and online game inspire the ubiquitous demands [1,2]. According to the statistics, numerous data transmissions are proceeding in residents or hot-spot areas [3]. Consequently, the desired quality of service (QoS) is accelerating the progress of hyper dense networks (HDN) in beyond 4G and 5G [4–6]. Heterogeneous and small cell networks (HetSNets), as one of the options for HDN, can support higher system throughput by introducing small cells (such as femtocells and picocells) [7,8]. The HetSNets infrastructure has been illustrated in Fig. 1, and different small cells have corresponding backhaul ways and scope of services [9]. However, massive deployments of small cells increase the signaling overhead and complexity of mobility management in HetSNets [10], and tracking area (TA) planning is part of the issue in hybrid mobility managements [11].

The dense deployments of small cells have been found to cause the heavy signaling overhead for location tracking with conventional principles of TA planning [12,13]. Several novel algorithms are therefore proposed to replan TA efficiently. In [14], an automatic replanning of TA for long term evolution (LTE) networks

© ICST Institute for Computer Sciences, Social Informatics and Telecommunications Engineering 2017
J. Cheng et al. (Eds.): GameNets 2016, LNICST 174, pp. 45–54, 2017.
DOI: 10.1007/978-3-319-47509-7_5

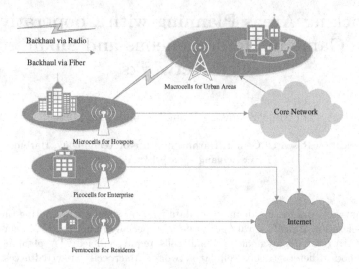

Fig. 1. Diagram of HetSNets infrastructure.

is presented via formulating the problem as a classical graph partitioning, which is solved by a multi-level graph partitioning algorithm. [15] presents an alternative method via modifying the handover decision and cell reselection, while considering the femtocells as groups for location tracking. Nevertheless, massive base stations consisting of the graph vertexes may bring the challenge of the algorithm proposed in [14], and forcing the users to stay in the macrocells as long as possible is not always efficient due to various user traffic and motion features [15].

Game theory has been widely used as a central tool for the design of future wireless and communication networks in recent years [16]. Game theory formulated interaction main incentive structures. It is a mathematical study of the theory and methods of the competitive nature of the phenomenon. Game theory considers the game to predict individual behavior and actual behavior, and to study their optimization strategy. Biologists use game theory to understand and predict the evolution of some of the results. Cooperative games encompass coalitional games that describe the formation of cooperating groups of players, referred to as coalitions. In [17], it utilizes coalitional games to detect communities in social networks. Inspired by [17], this paper considers TA planning as a detection of similar communities based on cooperative games.

2   System Model

Small cells are low-power access nodes, working on both the authorization and non-licensed spectrum. The coverage is from 10 m to 200 m, compared to the coverage of the macrocell, which can reach several kilometers. Mobile operators are worried about the growth of data traffic, and many operators believe that

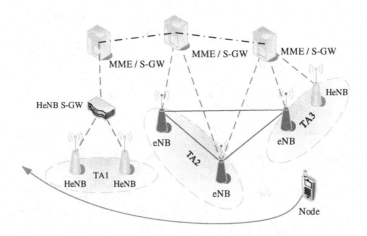

Fig. 2. Diagram of tracking area configuration.

mobile data is a good way to bypass the efficient use of radio spectrum resources. Data distribution is an important component that many operators take seriously. Small cells are an effective way to manage LTE -A spectrum instead of just using the macrocell. Small cells can be used both indoors and outdoors. Mobile operators is eager to extend the use of small cell cover range and improve the network capacity. In hot-spot areas, mobile operators can offload 80 % of the traffic. The report predicts that by 2015, 48% of mobile data traffic is streaming out from the macrocell. No single technology can rule the data distribution. It also believes that operators can discover new profit by small cell growth. When the registered user enters a femtocell zone, the network can learn the location information. With the obtaining permission, the location information can be updated immediately to social networks.

4G LTE networks defines the traditional method of location management as TA, which has its own identity called tracking area identity (TAI). One TA without overlapping is comprised of a group of continuous base stations. In each TA, the core network of LTE sends the paging request (PR) to idle users while the calling arrives, and the mobility management entity (MME) is responsible for location updating (LU) while the users move out of the TA. Consequently, the upper bound of TA is limited by the maximum paging load that the MME supports, and the lower bound is determined by signaling overheads that the users go across the TAs. The basic principles of TA planning are to balance PR load and LU signaling overhead, which are namely unique frequencies, MME, continuous areas, and topography oriented. The principle can avoid the channel congestion of PR, while reducing the LU signaling overhead. As shown in Fig. 2, the Femtocells in the same area may belong to different MME. If the Femtocell and eNB are considered as single TA, the backhaul bandwidth is a limitation of the fast transmission of paging signalling. Moreover, if the coexisted Femtocells are considered as single TA, the limited coverage of this TA may cause

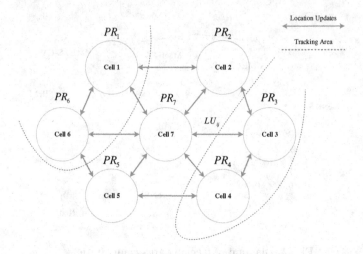

Fig. 3. TA planning as a graph modeling.

massive location updating signalling due to the frequent moving out of the TA. Therefore, the conventional TA replanning is not quite appropriate for HetSNets. The application of the corresponding location area concept, this area is called tracking area location. EPC for the UE is an idle state and connection status, and have their registered TA management, exchanging the registration information of the UE. EPC also deals with TA changed. Tracking area update (TAU) can tell EPC that UE is available, or handover between cellulars. Tracking area identity (TAI) is not in the list of UE. When TA is engaged in registration, it is necessary to perform TAU process.

Figure 3 shows a graph model $G(V, E)$ of TA planning, whose vertices V and edges E represent the base stations and adjacency of networks, respectively. The weight of vertex PR_i is the number of paging request in cell i, and the weight of edge LU_{ij} is the number of idle users, who move from cell i to cell j. We define n partitions of $G(V, E)$ as $G_1, G_2, \cdots G_n$, therefore, the optimal TA planning can be modeled as

$$\text{Min} \sum_{(i,j)\in(V_1,\cdots,V_n)} LU_{ij} \tag{1}$$

$$\text{s.t.} \sum_{i\in V_k} PR_i \leq C_{\max} \quad \forall k = 1 : n \tag{2}$$

where C_{max} is the capacity of the paging channel. In the following section, the solution of this problem is discussed.

3 Proposed TA Planning Algorithm Based on Cooperative Game

Before the cooperative game is introduced to solve the classical graph model presented in the previous section, we need to transform the expression of $G(V, E)$.

For all $LU_{ij} \geq 1$ and $LU_{ij} \in \mathbf{Z}$, expand $G(V, E)$ to $G'(V, E)$ via generating LU_{ij} vertices of v_i and v_j themselves, and making the new v_i' and v_j' connected. Namely, the original graph is extended by self-replicating based on value of LU_{ij}. Therefore, the new graph can be obtained by $G(V, E) = G'(V, E)$.

For the new $G(V, E)$, if v_i and v_j $(i, j \in N)$ have connections, we define $e_{ij} = 1$, or $e_{ij} = 0$. Therefore, the community detection model based on cooperative game can be expressed by CG $= (N, \text{Eigen})$, shown as

$$\text{Eigen}(S) = \begin{cases} 0 & S \subseteq N, |S| = 1, or, S = \varPhi \\ \sum\limits_{i \in S} \sum\limits_{j \in S, j \neq 1} \frac{e_{ij}}{d(i)} & S \subseteq N, |S| \geq 2, d(i) \neq 0 \end{cases} \quad (3)$$

where $d(i) = \sum\limits_{j \in N} e_{ij}$ is the degree of v_i and $\text{Eigen}(S)$ represents the benefit corresponding to the sum value of all edges [17]. So the Shapley value SH_{Eigen} can be calculated as

$$SH_{\text{Eigen}}(S, i) = SH_{\text{Eigen}}(S_1, i) + \frac{1}{2} \sum\limits_{j \in S_2} \left(\frac{e_{ij}}{d(i)} + \frac{e_{ji}}{d(j)} \right), i \in S_1 \quad (4)$$

$$SH_{\text{Eigen}}(S, j) = SH_{\text{Eigen}}(S_2, j) + \frac{1}{2} \sum\limits_{i \in S_1} \left(\frac{e_{ij}}{d(i)} + \frac{e_{ji}}{d(j)} \right), j \in S_2 \quad (5)$$

where $\forall S_1, S_2 \subseteq N$, $S = S_1 + S_2$. The detailed solution algorithm is shown as below:

Algorithm 1. CG procedure

---
get the network vertices $N = 1, 2, \cdots, n$
get the network edges $E = (e_{ij})_{n \times n}, i, j \in N$
suppose the level number $l = 1$
suppose the set of coalitions $S^l = \{\varPhi\}$
for $i = 1$ to n **do**
  get the ith coalition from S^l $S_i^l = \{i\}, S^l = S^l \cup \{S_i^l\}$
end for
while $|S^l| > 1$ **do**
  $l = l + 1$
  for all $x \in S_i^{l-1}$ such that $S_j^{l-1}, S_k^{l-1} \in S^{l-1}, j \neq k$, or $S_k^{l-1} = \varPhi$ **do**
    if $SH_e \left(S_i^{l-1} \cup S_j^{l-1}, x \right) \geq SH_e \left(S_i^{l-1} \cup S_k^{l-1}, x \right)$ **then**
      $S_r = S_i^{l-1} \cup S_j^{l-1}$
      $S^l = S^{l-1} - \{S_i^{l-1}\} - \{S_j^{l-1}\}$
      $S^l = S^l + \{S_r\}$
    end if
  end for
end while
get communities from $S^{l'}$

---

4   Performance Evaluation

For definiteness and without loss of generality, this paper considers a two-tier HetSNet, which is comprised of macrocell and femtocell operated with open access [18]. Additionally, macrocell is modeled as a hexagonal with three sectors depending on the carrier's deployment. Femtocell is assumed for a random distribution that follows a Poisson Point Process (PPP) based on stochastic geometry theory [19]. The SINR layout in the integration of macro cells and small cells is shown in Fig. 4.

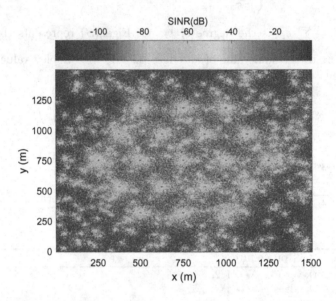

Fig. 4. SINR layout in the integration of macro cells and small cells.

According to 3GPP, the particular simulation parameters are summarised in Table 1. For the traffic type, the calling arrival rate λ is subject to a homogeneous Poisson process, and the mean holding time is 90 ms. For the mobility model, this paper introduces a opportunistic one that represents the human motion features in hot-spot area [20]. To illustrate the advantage of the proposed algorithm, two representative methods [14,15] are introduced as comparisons, which are short for TORIL 2013 and YU 2013 respectively.

The performance metric of TA planning is the total signalling overhead C_{total}, which is calculated by the sum of TA updating and PR, shown as

$$C_{\text{total}} = p\{\text{paging}\} \cdot \bar{N}_{\text{cells}} \cdot c_{\text{p}} + p\{\text{TAU}\} \cdot c_{\text{tau}} \tag{6}$$

where $p\{\text{paging}\}$ is the arriving rate of paging in one time slot, $\bar{N}_{\text{cells}}$ is the average number of cells in the TA list, c_p is the signalling overhead of every

Table 1. Simulation parameters

Parameter	Value
Carrier frequency/system bandwidth	2.0 (GHz)/10 (MHz)
UE distribution/speed	Uniform/30 (km/h)
Channel model	Typical urban (6 rays)
Transmit power of macro/femto	46 (dBm)/20 (dBm)
Path loss model (Macro)	$128.1 + 37.6\log10(R)$ (dB)
Path loss model (Femto)	$127 + 30\log10(R)$ (dB)
Shadowing standard deviation	Macro 8 (dB), Femto 4 (dB)
Macro/femto antenna gain	14 (dBi)/5 (dBi)
Macro/femto noise figure	5 (dB)/8 (dB)
A3 offset/TTT	3 (dB)/160 (ms)
Handover decision delay	50 (ms)
Handover execution time	40 (ms)

paging operation, and $p\{\text{TAU}\}$ is the probability of TA updating with the cost of c_{tau}. Generally, we make $\lambda = p\{\text{paging}\} = 0.05$, $c_{\text{tau}} = 10 \cdot c_p$.

Figure 5 shows the normalized signaling cost with various rate of femtocelll residence. Due to hyper dense deployments of femtocells in the given circumstance, the probability grows slowly as the increasing rate of femtocelll residence. It is clear that the method (YU 2013) via forcing the user to attach the macrocell as much as possible demonstrates the best signaling cost performance, and the proposed algorithm with cooperative game shows a better performance compared to TORIL 2013.

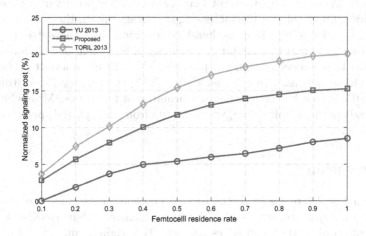

Fig. 5. The normalized signalling overhead with various rate of femtocelll residence.

The deployment of small cells is to improve the transmissions rate and offload the traffic from macrocells in hot-spots, so it is beneficial for users to attach small cells as much as possible. $p(k)$ is defined as the access probability to femtocells, where k is the sample time. The utilization of femtocells can be expressed as

$$U_{\text{small-cell}} = \sum_{k=1}^{n} P[k] \left(\sum_{j=1}^{k} p[j]/k \right) \tag{7}$$

where n is the sample time when the user is out of service by femtocells.

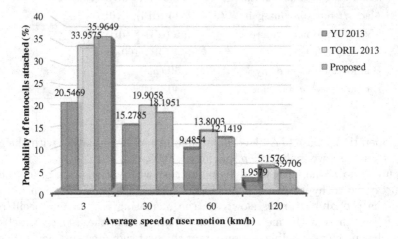

Fig. 6. Probability of femtocells attached with various speed of user motions.

Figure 6 describes probability of femtocells attached with various speed of user motions. Obviously, as the increasing of average speed, the utilization rate of femtocells is reducing, since the handover decision shall guarantee a specific successful access to networks for limited coverage of femtocells. Due to forcing the user to stay in macrocells as long as possible, YU 2013 has a lowest probability of femtocells attachment. In contrast to YU 2013, the proposed algorithm and TORIL 2013 have a better utilizing performance of femtocells. Meanwhile, the two methods perform quite close, which results from the optimal springboard of TA planning.

5 Conclusion

In this paper, a TA planning method based on cooperative game was proposed to reduce the signalling overhead of location management in HetSNets. After self replication of vertices and edges based on the paging requests and location updates, the modeled graph will be classified into new communities, which represent a planning of TA. Simulation results showed that the proposed algorithm reduces the signalling overhead while maintaining the utilization proportion of femtocells. In the future, the proposed method will be performed a detailed analysis on human mobility features that affect TA planning in various deployment scenarios.

Acknowledgment. This work has been sponsored by National Natural Science Foundation of China (No. 61101125 and 61571316), and the China Scholarship Council (No. 201406120100). Meanwhile, the authors would like to thank anonymous for improving the quality of this paper.

References

1. Zhang, H., Chu, X., Guo, W., Wang, S.: Coexistence of wi-fi and heterogeneous small cell networks sharing unlicensed spectrum. IEEE Commun. Mag. **53**(3), 158–164 (2015)
2. Zhang, H., Jiang, C., Beaulieu, N.C., Chu, X., Wang, X., Quek, T.Q.: Resource allocation for cognitive small cell networks: a cooperative bargaining game theoretic approach. IEEE Trans. Wireless Commun. **14**(6), 3481–3493 (2015)
3. Ning, L., Wang, Z., Guo, Q., Zhang, H.: Dynamic PCI assignment in two-tier networks based on cell activity prediction. Electronics Letters, efirst (2016). doi:10.1049/el.2016.0048
4. Andrews, J.G., Claussen, H., Dohler, M., Rangan, S., Reed, M.C.: Femtocells: Past, present, and future. IEEE J. Selected Areas Commun. **30**(3), 497–508 (2012)
5. Andrews, J.G.: Seven ways that hetnets are a cellular paradigm shift. IEEE Commun. Mag. **51**(3), 136–144 (2013)
6. Andrews, J.G., Buzzi, S., Choi, W., Hanly, S.V., Lozano, A., Soong, A.C., Zhang, J.C.: What will 5G be? IEEE J. Selected Areas Commun. **32**(6), 1065–1082 (2014)
7. Bangerter, B., Talwar, S., Arefi, R., Stewart, K.: Networks and devices for the 5G Era. IEEE Commun. Mag. **52**(2), 90–96 (2014)
8. Fortes, S., Aguilar-García, A., Barco, R., Barba, F., Fernández-luque, J., Fernández-Durán, A.: Management architecture for location-aware self-organizing lte/lte-a small cell networks. IEEE Commun. Mag. **53**(1), 294–302 (2015)
9. Zhang, H., Jiang, C., Rose Qingyang, H., Qian, Y.: Self-organization in disaster resilient heterogeneous small cell networks. IEEE Network preprint arXiv:1505.03209 (2015)
10. Zhang, H., Jiang, C., Cheng, J.: Cooperative interference mitigation and handover management for heterogeneous cloud small cell networks. IEEE Wireless Commun. **22**(3), 92–99 (2015)
11. Ferragut, J., Mangues-Bafalluy, J.: A self-organized tracking area list mechanism for large-scale networks of femtocells. In: IEEE International Conference on Communications (ICC), pp. 5129–5134. IEEE (2012)
12. Chatzikokolakis, K., Kaloxylos, A., Spapis, P., Alonistioti, N., Zhou, C., Eichinger, J., Bulakci, O.: A survey of location management mechanisms and an evaluation of their applicability for 5G cellular networks. Recent Adv. Commun. Networking Technol. **3**(2), 106–116 (2014)
13. Huai-Lei, F., Lin, P., Lin, Y.-B.: Reducing signaling overhead for femtocell/macrocell networks. IEEE Trans. Mobile Comput. **12**(8), 1587–1597 (2013)
14. Toril, M., Luna-Ramírez, S., Wille, V.: Automatic replanning of tracking areas in cellular networks. IEEE Trans. Vehicular Technol. **62**(5), 2005–2013 (2013)

15. Yifan, Y., Daqing, G.: The cost efficient location management in the lte pico-cell/macrocell network. IEEE Commun. Lett. **17**(5), 904–907 (2013)
16. Han, Z.: Game theory in wireless and communication networks: theory, models, and applications. Cambridge University Press (2012)
17. Zhou, L., Cheng, C., Lü, K., Chen, H.: Using coalitional games to detect communities in social networks. In: Wang, J., Xiong, H., Ishikawa, Y., Xu, J., Zhou, J. (eds.) WAIM 2013. LNCS, vol. 7923, pp. 326–331. Springer, Heidelberg (2013). doi:10.1007/978-3-642-38562-9_33
18. Chu, X., López-Pérez, D., Yang, Y., Gunnarsson, F.: Heterogeneous Cellular Networks: Theory Simulation and Deployment. Cambridge University Press, Cambridge (2013)
19. ElSawy, H., Hossain, E., Haenggi, M.: Stochastic geometry for modeling, analysis, and design of multi-tier and cognitive cellular wireless networks: A survey. IEEE Commun. Surv. Tutorials **15**(3), 996–1019 (2013)
20. Ning, L., Wang, Z., Guo, Q.: Preferred route indoor mobility model for heterogeneous networks. IEEE Commun. Lett. **18**(5), 821–824 (2014)

A Mechanism Design Approach for Influence Maximization

Michael Levet[1,2]([✉]) and Siddharth Krishnan[3,4]

[1] Department of Computer Science and Engineering,
University of South Carolina, Columbia, USA
mlevet@email.sc.edu
[2] Department of Mathematics, University of South Carolina, Columbia, USA
[3] Discovery Analytics Center, Virginia Tech, Blacksburg, USA
[4] Department of Computer Science, Virginia Tech, Blacksburg, USA
siddkris@cs.vt.edu

Abstract. With the proliferation of online social networks (OSNs), the characterization of diffusion processes and influence maximization over such processes is a problem of relevance and importance. Although several algorithmic frameworks for identifying influential nodes exist in literature, there is a paucity of literature in the setting of competitive influence. In this paper, we present a novel mechanism design approach to study the initial seeding problem where the agents, represented by vertices in the social network, are economically rational. The principals compete for influence in the network by setting price and incentives to illicit high degree initial subscribers, which in turn profit by infecting their neighbors. We restrict attention to equilibrium strategies and comparative statics for the agents.

Keywords: Social networks · Mechanism design · Influence maximization · Agents · Game theory

1 Introduction

The widespread adoption of online social media and networks, blogs, and internet shopping has transformed the web into a rich complex network structure. The web has therefore become a medium for diffusion of influence and information propagation. These spread processes have several applications including

Supported by the Intelligence Advanced Research Projects Activity (IARPA) via Department of Interior National Business Center (DoI/NBC) contract number D12PC000337, The US Government is authorized to reproduce and distribute reprints of this work for Governmental purposes notwithstanding any copyright annotation thereon. Disclaimer: The views and conclusions contained herein are those of the authors and should not be interpreted as necessarily representing the official policies or endorsements, either expressed or implied, of IARPA, DoI/NBC, DTRA, or the US Government.

© ICST Institute for Computer Sciences, Social Informatics and Telecommunications Engineering 2017
J. Cheng et al. (Eds.): GameNets 2016, LNICST 174, pp. 55–62, 2017.
DOI: 10.1007/978-3-319-47509-7_6

adoption of innovations [9], viral marketing [1], spread of rumors [3], and online recruitment. Information cascades resulting from such spreading processes have been widely observed in online social networks such as Facebook and Twitter. The algorithmic characterization of such processes is presented in two widely accepted models - the Linear Threshold and Independent Cascade [5,7]. The influence maximization process over these two models was first presented in [8], where a principal seeks to find the optimal set of nodes to seed. Recent literature has also focused on a Competitive Contagion model in which two or more principals compete for influence in the network [2,6] by seeding nodes in the network. An uninfected node changes state based on a stochastic process [6], rather than economic motivators, which is a major limitation in most algorithms that identify nodes of influence and agents of rapid propagation. We aim to bridge that gap in this contribution. In our work, we cast the problem as one of a referral game (described later) over a social network. The primary contribution of this work is the construction equilibrium strategies for the agents. Based on the importance of high-degree nodes in propagating influence [10], our second contribution provides conditions under which an agent will prefer seeding from one principal over another based on principal incentives and the agent's neighborhood size. Additionally, this model does not suffer from issues of complexity that plague the algorithmic approach [8,10].

2   Model

We examine a dynamic game of imperfect information, which we call the *referral game*, over a social network. The players of this game are partitioned into two disjoint, finite sets. The first set $\mathcal{P}$ consists of the principals, and the second set $\mathcal{A}$ consists of the agents. The set of players in this game is $N := \mathcal{P} \dot\cup \mathcal{A}$. The social network is represented by an undirected graph $G(V, E)$ where $V(G) := \mathcal{A}$. Two agents $i, j \in \mathcal{A}$ are related if $ij \in E(G)$. We postulate that each agent is only aware of its neighborhood set. That is, each agent v believes that with probability 1 the social network is $K_{1,|N(v)|}$, where $N(v)$ denotes the open neighborhood of v. Furthermore, we postulate that each principal is only aware that G exists, but not of any properties of G.

Each principal $p \in \mathcal{P}$ provides a membership option for each agent at price $c_p \in \mathbb{R}_+$. This membership provides a valuation $v_p \in \mathbb{R}_+$ to the member, which is exogenously given for principal p. For every principal p, both the price c_p and valuation v_p are public information. Initially, each agent does not hold a membership, and an agent can hold at most one membership which cannot be revoked or changed later.

Definition 1 (Infection). *An agent v is said to be infected if v holds a membership from some principal i. The agent v is said to be uninfected otherwise.*

The principals each seek to maximize influence over a graph. Define the membership function mem : $\mathcal{A} \to \mathcal{P} \cup \{\mathcal{U}\}$ which takes an agent and returns the principal from whom the agent is infected, or $\mathcal{U}$ if the agent is uninfected.

Each principal i has utility function: $u_i(\mathcal{A}) = |\mathrm{mem}^{-1}(i)|$. In order to maximize influence, each principal sets its price and incentives. Each principal's strategy set is $S := \mathbb{R}_+ \times [0,1] \times \mathbb{R}_+$. Principal i's strategy $(c_i, \alpha_i, \beta_i) \in S_i$ is interpreted as follows: c_i is the principal's price for infection, α_i is the price discount for agents who subscribe at $T = 0$ (that is, each agent pays $\alpha_i c_i$), and β_i is the amount an agent receives for each additional agent it refers. Furthermore, agents who are referred to principal i at some time $T > 0$ also receive β_i (but must still pay the undiscounted price c_i). Each principal's strategy constitutes a mechanism with which to infect willing agents. The collection of these mechanisms induces the referral game. For the purpose of this paper, principals' strategies are taken as fixed. We restrict attention to the comparative statics with respect to the agents rather than the principals' strategies.

The game operates in discrete time steps starting at time $T = 0$. Prior to the start of the game, each principal fixes its strategy. At $T = 0$, any agent may become infected by at most one principal. After this initial joining period, uninfected agents can only become infected by principal i through referral from an agent already infected by i. An uninfected agent can only receive such a referral from one of its neighbors. At each discrete time $T > 0$, each infected agent may submit proposals simultaneously to any subset of its uninfected neighbors. The uninfected nodes that received proposals may accept at most one of the referrals or remain uninfected.

We now define each agent's utility function. Denote $\mathcal{U}$ as the option of remaining uninfected. Each agent v has the partial utility function of the form:

$$u_v : (\mathcal{P} \cup \{\mathcal{U}\}) \times (N(v) \cup \mathcal{P} \cup \{\emptyset\}) \times 2^{N(v)} \to \mathbb{R}_+ \qquad (1)$$

The component $(\mathcal{P} \cup \{\mathcal{U}\})$ denotes the principal by whom v is infected, or whether v is uninfected. The component $N(v) \cup \mathcal{P} \cup \{\emptyset\}$ describes the player referring v. If the player is a principal, this indicates that v subscribed to the principal at time $T = 0$. The $\emptyset$ option denotes that no such referral has been made, and v is uninfected. Finally, the element from $2^{N(v)}$ denotes the set of neighbors which v successfully referred. We define the following cases:

1. $u_v(\mathcal{U}, \emptyset, \emptyset) = 0$. That is, an agent experiences no utility for remaining uninfected.
2. $u_v(i, i, s) = v_i - \alpha_i c_i + |s|\beta_i$ for any $i \in \mathcal{P}$ and any $s \in 2^{N(v)}$. This case indicates that agent v subscribed at time $T = 0$ to principal i.
3. $u_v(i, j, s) = v_i - c_i + (|s| + 1)\beta_i$ for any $i \in \mathcal{P}$, any $j \in N(v)$, and any $s \in 2^{N(v) \setminus \{j\}}$. This case indicates that agent v was referred by its neighbor j to principal i at time $T > 0$, and then v referred an additional $|s|$ of its uninfected neighbors.

3   Analysis

The *referral game* is a dynamic game of imperfect information. The solution concept we use is the Perfect Bayesian Equilibrium. We construct a Perfect

Bayesian Equilibrium for this game using backward induction. At every time $T > 0$, each infected agent can only attempt to infect uninfected neighbors. Uninfected neighbors can only accept or reject referrals when received. We first show that an agent infected at time T need only attempt to infect its neighbors at time $T + 1$. This determines the Nash equilibria for terminal subgames.

Proposition 1. *Let x be an agent infected by principal i. Let $N_u(x)$ denote the set of uninfected neighbors of x. Then x will propose to each neighbor $y \in N_u(x)$ exactly once.*

Proof. Fix $y \in N_u(x)$. It is a weakly dominant strategy for x to propose to y. If y accepts a proposal from one of its infected neighbors, then we are done. Suppose instead at time $T_k > 0$, y rejects each of its infected neighbors' proposals. Recall that under y's belief system, the social network is $K_{1,|N(y)|}$, with y at the center. We thus have $\mathbb{E}[u_y(i, x, N_u(y))] = v_i - c_i + (|N_u(y)| + 1)\beta_i < 0$. As no infected agent can become uninfected, $|N_u(y)|$ is non-increasing as the game progresses. So $\mathbb{E}[u_y(i, x, N_u(y))] < 0$ for every time $T > T_k$.   □

We now use Proposition 2 to show that the game is finite, which implies the existence of an equilibrium [4].

Proposition 2. *Denote the set of agent states $K = \mathcal{P} \cup \{\mathcal{U}\}$, where u denotes a vertex remaining uninfected and each element $i \in \mathcal{P}$ denotes infection by principal i. The vertex states converge to a steady state equilibrium in $K^{|V|}$. Furthermore, this equilibrium is reached in $O(|V|)$ time steps.*

Proof. Once an agent becomes infected, its state is fixed. From Proposition 2, an agent v that remains uninfected at time $T = 0$ receives at most $|N(v)|$ referrals, of which v can select at most one. It follows that the vertex states converge to a steady state equilibrium in $K^{|V|}$. The bound is tight, as in the case of the graph G is a path on $|V|$ vertices where one endpoint agent becomes infected at $T = 0$. Then at most one additional agent is infected by referral at each subsequent time step, implying that the vertex states will converge to equilibrium in at most $|V|$ time steps.   □

Corollary 1. *There exists a Perfect Bayesian Equilibrium [4].*

The following proposition will construct an explicit Perfect Bayesian Equilibrium in mixed strategies for the case when $G \cong K_2$. This result will be used to construct the Perfect Bayesian Equilibrium for the general case.

Proposition 3. *Suppose the social network $G \cong K_2$. Let $j = \arg\max_{i \in \mathcal{P}} v_i - \alpha_i c_i$. Define: $Q_1 = \{i \in \mathcal{P} : v_i - c_i + \beta_i < v_j - \alpha_j c_j\}$ and let $Q_2 = \mathcal{P} \backslash Q_1$. Then there exists a symmetric Perfect Bayesian Equilibrium in mixed strategies, where each agent considers at most three principals.*

Proof. As this game is symmetric, there exists a Perfect Bayesian Equilibrium in mixed strategies where each agent employs the same strategy. [4] Consider the following cases:

Case 1: Suppose $Q_1 = \mathcal{P}$ and $Q_2 = \emptyset$. Then no player benefits from infection by referral. Therefore, each agent will always choose infection from j at time $T = 0$ if and only if $u(j, j, \emptyset) \geq 0$. Otherwise, each agent will remain uninfected. By construction, this strategy maximizes each player's payoff.

For the rest of the proof, assume there exists at least one principal x such that $v_x - c_x + \beta_x \geq 0$. For if no such principal x exists, then each agent will remain uninfected as the equilibrium strategy.

Case 2: Suppose $Q_1 = \emptyset$ and $Q_2 = \mathcal{P}$. Define:

$$k = \arg\max_{i \in \mathcal{P}} v_i - \alpha_i c_i + \beta_i \tag{2}$$

$$\text{s.t. } v_i - c_i + \beta_i \geq 0 \tag{3}$$

By construction of Q_2, such a k always exists. The game has exactly one stage if and only if both agents play the same strategy of either subscribing to principal k or remaining uninfected at time step $T = 0$. Otherwise, the uninfected agent at $T = 0$ will accept the referral at time $T = 1$. We now solve for the symmetric mixed strategies equilibrium. Suppose player 2 chooses infection at time $T = 0$ with probability p and chooses to remain uninfected at $T = 0$ with probability $1 - p$. Then player 1's expected payoffs from becoming infected and remaining uninfected at $T = 0$ respectively are:

$$\mathbb{E}[u_1(k, k, N(v_1))] = p \cdot (v_k - \alpha_k c_k) + (1 - p) \cdot (v_k - \alpha_k c_k + \beta_k) \tag{4}$$

$$\mathbb{E}[u_1(k, v_2, \emptyset)] = p \cdot (v_k - c_k + \beta_k) \tag{5}$$

Setting $\mathbb{E}[u_1(k, k, N(v_1))] = \mathbb{E}[u_1(k, v_2, \emptyset)]$ and solving for p yields:

$$p = \frac{v_k - \alpha_k c_k + \beta_k}{v_k - c_k + 2\beta_k} \tag{6}$$

Case 3: Suppose $Q_1 \neq \emptyset$ and $Q_2 \neq \emptyset$. Define k as in case 2. If $k \in Q_2$, then we reduce to case 2. Otherwise, suppose $k \in Q_1$. Define: $m = \arg\max_{i \in Q_2} v_i - \alpha_i c_i + \beta_i$. That is, m is an agent's preferred principal for infection at time $T = 0$ such that its neighbor will accept the proposal if uninfected. From the definition of j and the fact that $k \in Q_1$, we have $v_i - c_i + \beta_i \geq 0$ for all $i \in Q_2$. By construction, $v_k - \alpha_k c_k + \beta_k > v_m - \alpha_m c_m + \beta_m$. Observe the strategy of choosing infection from principal k at time $T = 0$ weakly dominates infection from any of the other principals of $Q_1 \backslash \{j\}$ at $T = 0$. Similarly, the strategy of choosing infection from principal m at time $T = 0$ weakly dominates infection to any other principal in $Q_2 \backslash \{j\}$ at time $T = 0$. We are thus left with four viable strategies for each agent: choose infection from principal $p \in \{m, j, k\}$ at $T = 0$ and then propose to its uninfected neighbor; and remain uninfected at $T = 0$, accepting any proposal where the expected payoff is non-negative. We now solve for a symmetric mixed strategies equilibrium.

Agent i mixes his strategies such that agent $-i$ is indifferent between the four viable strategies. Let $p_j, p_k,$ and p_m denote the frequencies in which each player at time $T = 0$ chooses infection from princiapls j, k and m respectively; and let p_u denote the frequency in which each player chooses to initiallly remain uninfected. For initially subscribing to a principal $y \in \{i, j, k\}$, each agent has the expected payoff:

$$\mathbb{E}[u_i(y, y)] = (1 - p_u) \cdot u(y, y, \emptyset) + p_u \cdot u(y, y, N(i)) = v_y - \alpha_y c_y + p_u \beta_y \quad (7)$$

And for opting to remain uninfected at time $T = 0$, each agent has the expected payoff:

$$\mathbb{E}[u_i(\mathcal{U})] = \sum_{x \in \{j,k,m\}} p_x \cdot u_i(x, -i, \emptyset) = \sum_{x \in \{j,k,m\}} p_x \cdot (v_x - c_x + \beta_x) \quad (8)$$

We solve for the mixed strategies equilibrium by setting (7) equal to (8), where we consider (8) for each $y \in \{i, j, k\}$. The following linear program yields such a mixed strategy equilibrium, with constraitnt (10) denoting the condition that (7) and (8) are equal in equilibrium.

$$\max_{p \in \Delta} \sum_{x \in \{j,k,m\}} p_x \cdot (v_x - c_x + \beta_x) \text{ s.t.} \quad (9)$$

$$\sum_{x \in \{j,k,m\}} p_x \cdot (v_x - c_x + \beta_x) = v_i - \alpha_i c_i + p_u \beta_i; \ \forall i \in \{j, k, m\} \quad (10)$$

□

Under an agent v's belief system, each of its neighbors believe $G \cong K_2$. As v believes $G \cong K_{1,|N(v)|}$, v believes each of its neighbors behaves independently and symmetrically in equilibrium. We use the mixed strategies equilibrium from Proposition 3 to construct v's equilibrium strategy at $T = 0$ based on $\mathbb{E}[\|N_u(v)\|]$.

Proposition 4. *Let $n \in \mathbb{Z}_{++}$ and suppose the social network $G \cong K_{1,n}$. Define $j, Q_1, Q_2,$ and p^* as in Proposition 3. Denote p_u^* to be the component of p^* corresponding to remaining uninfected at time $T = 0$. Let v be the center vertex of G. Define: $m_1 = v_j - \alpha_j c_j$ and $m_2 = \max_{i \in \mathcal{P}} v_i - \alpha_i c_i + |N(v)| \cdot p_u^* \beta_i$. In equilibrium, agent v's expected utility is: $\max\{m_1, m_2, 0\}$.*

Proof. Recall that each node is only aware of its neighbors. We assume that each vertex in $N(v)$ plays the equilibrium strategy described in Proposition 3. Let X be the binomial random variable associated with the number of vertices in $N(v)$ which remain uninfected at time $T = 0$. Agent v seeks to maximize his expected utility. Suppose $p_u^* > 0$ and suppose there exists such a principal i such that $v_i - c_i + \beta_i \geq 0$. Now suppose v plays the strategy of subscribing to some principal i at time $T = 0$, then proposing to each uninfected neighbor at time $T = 1$. Then: $\max_{i \in \mathcal{P}} \mathbb{E}[u_v(i, i, N(v))] = m_2$. If agent v instead opts not to propose to each of its neighbors at time $T = 1$, then his maximum utility is $m_1 = v_j - \alpha_j c_j$, which is obtained by subscribing to agent j at time $T = 0$. Agent v's third option is to remain uninfected at time $T = 0$. Agent v chooses the best of these three strategies in equilibrium. □

The following theorem specifies and verifies a Perfect Bayesian Equilibrium.

Theorem 1. *Let G be a simple, undirected graph. Suppose each agent v plays the strategy specified by Proposition 4. This constitutes a Perfect Bayesian Equilibrium.*

Proof. Let v be an agent. For each time $T > 0$, v's strategy consists of proposing to each uninfected neighbor if v is infected; or if v is uninfected, it accepts a referall from its neighbor infected by principal j, which we denote x_j, if $x_j \in \arg\max_{i \in N(v)} u_v(i, x_i, N_u(v))$. By Proposition 1, this strategy induces a Nash equilibrium at each subgame for every $T > 0$. Recall that agent v believes with probability 1 that $G \cong K_{1,|N(v)|}$. It follows that the strategy at time $T = 0$ specified by Proposition 4 constitutes a Nash equilibrium. From this and Proposition 1, this strategy is sequentially rational. Consistency follows immediately from the fact that v believes $G \cong K_{1,|N(v)|}$ with probability 1. □

Finally, we examine the comparative statics, deriving sensitivity results for an agent's preferences for infection from a specific principal. We relate the size of an agent's neighborhood in the network to perturbations of the α and β parameter's in a principal's incentives package.

Theorem 2. *Let G be a graph, and let x be an agent. Suppose $\mathcal{P} = \{i, j\}$, $v_i = v_j$, and $c_i = c_j$, which we denote as v and c respectively. Then x strictly prefers infection from principal i rather than principal j at time $T = 0$ if one of the following conditions hold:*

1. *$\alpha_i \leq \alpha_j$, $\beta_i \geq \beta_j$ and at least one of the inequalities is strict.*
2. *$\alpha_i > \alpha_j$, $\beta_i > \beta_j$ and either: $\beta_i - (1 - \alpha_j)c > 0$; or $|N(x)| > \dfrac{(\alpha_i - \alpha_j)c}{\beta_i - \beta_j}$.*

Proof. If $\alpha_i \leq \alpha_j$ and $\beta_i \geq \beta_j$, then $u_x(i, i, S) \geq u_x(j, j, S)$ for every $S \subset N(x)$. So in this case, x prefers holding infection from principal i over principal j. Now suppose instead that $\alpha_i > \alpha_j$ and $\beta_i > \beta_j$. As x prefers infection from i over j at $T = 0$, it is necessary that under x's belief system, the following conditions must hold:

1. Under x's belief system, each $y \in N(x)$ prefers referral to i rather than joining j at $T = 0$.
2. Under x's belief system, each $y \in N_u(x)$ would accept referral to j. However, $u_x(i, i, N(x)) > u_x(j, j, N(x))$.

Condition one is equivalent to $\beta_i - c > -\alpha_j c$, which implies that $\beta_i - (1 - \alpha_j)c > 0$. Condition two is equivalent to $-\alpha_j c + |N(x)|\beta_j < -\alpha_i c + |N(x)|\beta_i$, which implies that $|N(x)| > \dfrac{(\alpha_i - \alpha_j) \cdot c}{\beta_i - \beta_j}$. □

4   Conclusion and Future Work

In this paper, we constructed equilibrium strategies for agents in the referral game under the assumption that the principals had no knowledge of the social network's structure. The natural extension of this work is to utilize these results to construct principals' equilibrium strategies. We propose designing a beliefs system regarding network connectivity for each principal, allowing for the continuation of the backward induction argument. Additionally, this model prohibits awareness of any mutual relations; while in most social situations, imperfect knowledge of mutual relations. We propose the problem of determining agent equilibrium strategies when every adjacent pair of agents i and j in the social network are aware of some $S \subset (N(i) \cap N(j))$ a priori. We are interested in the symmetric case, where related i and j are aware of the same $S \subset (N(i) \cap N(j))$, as well as the asymmetric case where i is aware of some $S_1 \subset (N(i) \cap N(j))$ and j is aware of some potentially different $S_2 \subset (N(i) \cap N(j))$.

Acknowledgments. We wish to thank Brendan Avent, Éva Czabarka, Stephen Fenner, and Alexander Matros for their helpful discussions and suggestions.

References

1. Domingos, P., Richardson, M.: Mining the network value of customers. In: Proceedings of the Seventh ACM SIGKDD International Conference on Knowledge Discovery and Data Mining, KDD 2001, pp. 57–66. ACM, New York (2001). ISBN 1-58113-391-X
2. Draief, M., Heidari, H., Kearns, M.: New models for competitive contagion (2014). http://www.aaai.org/ocs/index.php/AAAI/AAAI14/paper/view/8399
3. Friggeri, A., Adamic, L.A., Eckles, D., Cheng, J.: Rumor cascades. In: Proceedings of the Eighth International Conference on Weblogs and Social Media, ICWSM 2014, Ann Arbor, Michigan, USA, 1–4 June 2014 (2014). http://www.aaai.org/ocs/index.php/ICWSM/ICWSM14/paper/view/8122
4. Fudenberg, D., Tirole, J.: Game Theory. MIT Press, Cambridge (1991)
5. Goldenberg, J., Libai, B., Muller, E.: Talk of the network: a complex systems look at the underlying process of word-of-mouth. Market. Letters **12**(3), 211–223 (2001)
6. Goyal, S., Kearns, M.: Competitive contagion in networks. CoRR, abs/1110.6372 (2011). http://arxiv.org/abs/1110.6372
7. Granovetter, M.: Threshold models of collective behavior. Am. J. Sociol. **83**(6), 1420–1443 (1978)
8. Kempe, D., Kleinberg, J., Tardos, É.: Maximizing the spread of influence through a social network. In: Proceedings of the Ninth ACM SIGKDD International Conference on Knowledge Discovery and Data Mining, KDD 2003, pp. 137–146. ACM, New York (2003)
9. Rogers, E.M.: Diffusion of innovations, 5th edn. Free Press, New York (2003). ISBN 0-7432-2209-1, 978-0-7432-2209-9
10. Seeman, L., Singer, Y.: Adaptive seeding in social networks. In: Proceedings of the 2013 IEEE 54th Annual Symposium on Foundations of Computer Science, FOCS 2013, pp. 459–468. IEEE Computer Society, Washington, DC (2013). ISBN 978-0-7695-5135-7

Energy Efficient Channel Sharing and Power Optimization for Device-to-Device Networks

Zeguo Xi[1,2]([✉]), Xiangming Wen[1,2], Zhaoming Lu[1,2],
Yan Zeng[1,2], Zhiqun Hu[1,2], and Tao Lei[1,2]

[1] Beijing Key Laboratory of Network System Architecture and Convergence,
Beijing University of Posts and Telecommunications, Beijing 100876, China
xizeguo1@bupt.edu.cn
[2] Beijing Laboratory of Advanced Information Networks, Beijing 100876, China

Abstract. In device-to-device (D2D) networks, the system performance can be significantly improved with a well resource allocation scheme. In this paper, the issue of channel sharing and power allocation for device-to-device (D2D) communications underlaying cellular networks is considered. The users with the same service content are categorized into clusters, with clusters sharing the frequency of the uplink users. With this non-orthogonal frequency sharing, the energy efficiency of different type of users, i.e., the uplink users or the D2D users, is analysed. The energy-efficient resource sharing problem is further formulated into a non-transferable coalition formation game, and several related factors of the game is described. A distributed coalition formation game algorithm based on the merge and split rule is proposed. With numerical results, the effectiveness of the game model and the algorithm is demonstrated.

Keywords: Energy efficiency (EE) · Device-to-device (D2D) communication · Coalition formation game · Resource sharing

1 Introduction

The substantial increase of network elements and users explosive data traffic requirements is the inevitable trend of todays wireless network, which brings a serious challenge for network management and business delivery. In order to improve the service efficiency and satisfy the users' service quality, resource allocation in heterogeneous network has been fully studied. [1] describes a network architecture which combines cloud radio access network with small cells, while [2] maximize the total capacity of all femtocell users without ignoring the fairness and the spectrum sensing errors. [3] proposes a novel semidynamic clustering scheme based on affinity propagation for picocell to maximize users spectrum efficiency and throughput, and [4] introduces a network architecture where small cells use the same unlicensed spectrum that Wi-Fi systems operate in without affecting the performance of Wi-Fi systems. However, the energy efficiency of the network is ignored in most of the existing studies. The energy consumption

© ICST Institute for Computer Sciences, Social Informatics and Telecommunications Engineering 2017
J. Cheng et al. (Eds.): GameNets 2016, LNICST 174, pp. 63–75, 2017.
DOI: 10.1007/978-3-319-47509-7_7

is tightly coupled with the scale of users, and the diversity of user distribution not only leads to the heterogeneity of radio channel and the poor communication conditions for some users, but also seriously affects the energy efficiency on the network side and the battery life. Note that the convergence of service and content is one of the salient features of wireless networks. [5,6] have shown that users in the same access point often have the similar service content request, and geographically adjacent users may have a similar content request. With such similarity in service content, user collaboration based on D2D transmission can take the advantage of the heterogeneity of multi-user channels and improve the energy efficiency.

In such D2D transmission underlaying cellular networks, interference need to be carefully considered. Frequency allocation between the potential D2D clusters and uplink users is an crucial issue. Distributed resource allocation algorithms which are based on the reverse iterative combinatorial auction (ICA) game and the bisection method were proposed in [7,8]. However, the quality of service (QoS) provisioning issue is not considered and no close-form solution has been derived. Centralized resource allocation algorithms for optimizing the energy efficiency in the device-to-multidevice (D2MD) and D2D-cluster scenarios were explored in [9,10]. However the computational complexity is high and the signalling is increasing significantly with the number of user equipments (UEs), it's hardly for the base station to deliver the information to the user equipments within the channel coherent time in practical. In [11] an auction-based resource allocation algorithm was proposed to maximize the battery lifetime, but the energy efficiency of cellular UEs were neglected.

In this paper, a coalition formation game model is proposed for resource sharing in mobile D2D communications underlaying cellular networks. As a useful tool to model the complex interactions among users while accounting for the inherent benefit-cost trade-off in [12], coalition formation game theory can be well qualified to design the resource sharing scheme for D2D communications [13]. In particular, the proposed resource sharing scheme is more practical than the previous works.

The main contributions of this paper are as follows: (1) Different from previous works aim at one potential D2D pair or cluster [14,15], the proposed scheme is suitable for multiple potential D2D clusters and multiple uplink users, which is more general. (2) An novel energy efficiency equation for nonorthogonal D2D communications is proposed, both the spectrum utilization and QoS constraints are considered. (3) The resource sharing problem is modeled as a non-transferable coalition formation game. With the process of coalition formation and the resulting partition, the joint optimization of channel sharing and power optimization is addressed. Compared with [16,17], our coalition formation algorithm is distributed, which allows the users to adapt to the environmental changes. And the proposed scheme is more flexible for the data-requesting users than other schemes such as in [18].

The rest of this paper is organized as follows: Sect. 2 introduces the system model of the D2D communication underlaying cellular networks. In Sect. 3, the resource sharing problem is formulated as a non-transferable coalition formation game, and an algorithm is proposed for the game to obtain the stable coalition structure. In Sect. 4, the algorithm is validated with numerous simulations. Section 5 gives the conclusion.

2 Network Model

We consider a single-cell network, the radius of which is R and a base station (BS) is located at the center. There are N users distributed randomly in the network, communicating with the BS through the uplink channel, they called uplink users. Moreover, there are M users requesting the same business content, the popular content could be a live show or a hot video. The M users called data-requesting users. Since they all need the same data, the data could be relayed from one user to another. These users could be composed into several collaboration clusters, and every cluster has one cluster head which receives traffic data from the BS through long-range communication, and then distributes the data to the other users within the cluster by short-range communications. Note that the short-range communications are operated in the form of broadcast. Especially, the relay in the cluster will reuse the uplink of some uplink users. Several clusters may be built, and there will always be some data-requesting users not in any cluster. The BS will regard the independent data-requesting users as normal downlink users.

The clusters are constructed basing on the distance relationship, and not every data-requesting users can be in a cluster. Let d be the maximum distance of D2D link, if the distance between two data-requesting users is less than d, then each of them has one neighbor. The user with the most neighbors is selected as the cluster head, and it will form a cluster with its neighbors being the corresponding cluster tails. The next cluster will be formed from the rest data-requesting users, and the cluster formation process will continue until the leaving data-requesting users have no neighbor.

As illustrated in Fig. 1, uplink users U_1, U_2, U_3, U_4 are communicating with the BS using different frequency band with a bandwidth normalized to one. There are seven data-requesting users requesting the same service content, and they are divided into three parts CL_1, CL_2, S_1. Note that CL_1, CL_2 are clusters. The cluster head in CL_1 transmits the data to its tails using the uplink of U_1 and U_3, while the cluster head in CL_2 reuses the uplink of U_2 for the short-range communications. There is only one user in S_1, which means the user has no neighbor within d, so it communicate with the BS directly.

From the aspect of green communication, we take the EE of each transmission link as the performance metric for the users. For every user in the network, including the uplink users and the data-requesting users, the EE of them is defined as the ratio of the throughput and the total power of the user's link. More precisely, the EE of the uplink user is the energy efficiency of data sending, while the EE of the data-requesting user is the energy efficiency of data receiving. And the total power of one link is consisted of two parts: the power of the power

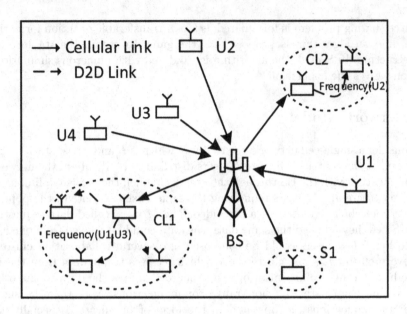

Fig. 1. A single-cell D2D underlaid cellular system

amplifier (PA) and the power consumed by circuit blocks of both the transmitter and the receiver. The corresponding functions are as follows:

$$EE = T/P_{link} \qquad (1)$$

$$P_{link} = P_{pa} + P_{ct} + P_{cr} \qquad (2)$$

$$P_{pa} = KP_t \qquad (3)$$

T is the throughput of one user, while P_{link} is the total power of the link. P_{pa} is the power of the PA, and P_t is the transmit power. The ratio of them is K, which is a value related to the modulation scheme. P_{ct}, P_{cr} respectively represents the power of the circuit blocks of the transmitter and the receiver, which remains the most basic function.

In order to obtain the EE of one user, two parameters must be ensured, the throughput of the user and the total power of the link. Generally, the first parameter can be set as the minimum required throughput, which also represents the QoS of the user. So the requirement of the user is satisfied. Once the throughput is setted, using Shannon's theorem the value of the received signal-to-interference-plus-noise ratio (SINR) can be calculated. After that, using the information of the interference and the noise, the transmit power of the link could be computed. Since the power of the circuit blocks of the transmitter and the receiver are known, the second parameter can be acquired.

The parameters of the uplink users, the cluster head users, and the data-requesting users with no neighbor are obtained using the following equations:

$$T = Blog_2(1 + SINR) \qquad (4)$$

$$SINR = P_t H/N \qquad (5)$$

B is the bandwidth of the link, P_t is the transmit power of the link. H is the channel gain from the transmitter to the receiver, and N is the noise. Since no user occupy the other users' resource, no interference considered.

For the tail users in the clusters and the uplink users who share their resource with the clusters, the calculation is complicated because of the mutual interference. The analysis is as follows.

First of all, the most remote tail user is studied. Assuming there is only one tail user in each cluster, and the user is the most remote one for the cluster head. Establish the equations of SINR and throughput of the single tail user and the resource-sharing uplink users, the corresponding transmit power can be calculated.

$$T_{tail} = B_t log_2(1 + SINR_{tail}) \qquad (6)$$

$$SINR_{tail} = P_{t,ch} H_{ch,tail} / (\sum_{i=1}^{M} P_{t,u} H_{u,tail} + N) \qquad (7)$$

$$T_{uplink} = B_u log_2(1 + SINR_{uplink}) \qquad (8)$$

$$SINR_{uplink} = P_{t,u} H_{u,B} / (P_{t,ch} H_{ch,B} + N) \qquad (9)$$

For the sake of simplicity, the most remote tail user is called single tail. T_{tail} is the throughput of the single tail, while the SINR of the single tail is $SINR_{tail}$. B_t is the bandwidth of the link of the single tail. $P_{t,ch}$ is the transmit power of the cluster head, and $H_{ch,tail}$ is the channel gain from the cluster head to the single tail. Similarly, the transmit power of the resource-sharing uplink user is $P_{t,u}$, and the channel gain from the uplink user to the single tail is $H_{u,tail}$, the product of the two value is the interference for the single user. There may be several uplink users sharing the resource with one cluster, so the interference is additive. For each resource-sharing uplink user, the throughput, bandwidth, and SINR are respectively denoted by T_{uplink}, B_u, and $SINR_{uplink}$. $H_{u,B}$ is the channel gain from the uplink user to the BS, and $H_{ch,B}$ is the channel gain from the cluster head to the BS. Using the equations, the value of $P_{t,ch}$ and $P_{t,u}$ can be ensured with the value of T_{tail} and T_{uplink}.

Secondly, the remaining cluster tails are discussed. As for the data-requesting user who is a tail user of one cluster but not the most remote one, both the two parameters depend on the most remote tail user. The transmit power of the link of the data-requesting user equals the transmit power from the cluster head to the most remote tail user. And then the throughput is obtained according to this power and the corresponding interference and noise. While the most remote tail user just meet the minimal throughput requirement, the throughput of the other tail users are higher.

$$SINR_{rt} = P_{t,ch} H_{ch,rt} / (\sum_{i=1}^{M} P_{t,u} H_{u,rt} + N) \qquad (10)$$

$$T_{rt} = B_{rt} log_2(1 + SINR_{rt}) \qquad (11)$$

$P_{t,ch}$ and $P_{t,u}$ have been calculated above. These data-requesting users are called the rest tails. And for each rest tail, the SINR and the channel gain from the cluster head to it are respectively denoted by $SINR_{rt}$ and $H_{ch,rt}$. $H_{u,rt}$ is the channel gain from the resource-sharing uplink user to the rest tail. The interference come from the same uplink users who share the resource with the most remote tail user, and the interference is additive. When $SINR_{rt}$ is obtained, we can use the bandwidth B_{rt} to get the throughput of this tail user T_{rt}.

The value of EE depends on the resource sharing result, so the problem of D2D resource allocation can be described as a process of the match between the uplink users and the D2D clusters.

3   Coalition Formation Game and the Solution

In order to solve the joint problem of uplink resource allocation and power management, the energy-efficient uplink resource sharing problem is modelled as a non-transferable coalition formation game. After weighing the benefits of the improvement of EE and the loss caused by mutual interference, the final match relationship will be obtained.

In the coalition formation game, several related factors are defined as:

Player: The set of game players is defined as X, which includes all of the uplink users and the data-requesting users. And they all attempt to merge with others to get the collection of coalitions more stable, and get all the EE improved.

Strategy: The collection of coalitions is defined as L, which describes the match relationship of the uplink users and the data-requesting users.

Utility: The characteristic function of a coalition is defined as CF, the value of which is based on the EE of the users in this coalition. Take coalition L_i as an example:

$$CF(L_i) = \{u_1(L_i), u_2(L_i), ..., u_r(L_i), ..., u_{|L_i|}(L_i)\} \tag{12}$$

$CF(L_i)$ is a vector, and $u_r(L_i)$ is the utility of player $r \in L_i$.

Since coalition L_i is obtained from the resource reused relationship between clusters and uplink users, the users in L_i have the following cases. In the first case, coalition L_i has only one user, which could either be an uplink user or a data-requesting user. None uplink resource will be reused and $u_r(L_i) = EE_r$. In the second case, coalition L_i contains one D2D cluster, which means no uplink users will share resource with this cluster. In this paper, D2D cluster is treated as an inseparable entity. The users in the D2D cluster will directly receive data from the BS and $u_r(L_i) = EE_r$. In the third case, coalition L_i consists of one D2D cluster and several uplink users who share their resource with the cluster. For an arbitrary user in the cluster, the utility is expressed as $u_r(L_i) = EE_r$. However, the utility expression of the uplink users can not be as simple as that. On the one hand they suffer from the interference from the D2D cluster, the energy efficiency would surely be reduced. On the other hand, they are inspired to share their resource. Therefore the utility must be adjusted. So the utility of the uplink users are defined as $u_r(L_i) = EE_r + \mu(u_{CL_i}(L_i) - u_{CL_i})$, where μ is a positive constant and CL_i is the cluster in coalition L_i. The second part of the utility is a compensate function, which indicates that the improvement of the utility of CL_i will be rewarded to the uplink users. Here the utility of CL_i is defined as the utility of the most remote tail user.

The case one uplink user share the resource with more than one cluster is not concerned, because the interference is too much. And the probability of a coalitions formation decrease with the increase in the number of uplink users in the coalition, for the costs limit the advantage. By well performing the uplink users' resource sharing, the utility of all users can be improved at the same time, and the new coalition structure are more beneficial.

The utility obtained by every user is related to the rest users in its coalition, and the coalition value cannot be arbitrarily apportioned among them, so the coalition formation game has non-transferable utility (NTU). Because of that, the Pareto Optimality can be used to judge the merits of collections of coalitions, which will be mentioned later. The increase in the cost depends on many factors, so the proposed coalition formation game is non-superadditive. Given one grand coalition which consists all users, there would be only one cluster reusing all the uplink resource. Not to mention the difficulty that all data-requesting users are distributed closely, the case all uplink users involve in the resource sharing is rarely seen. When the number of uplink users and data-requesting users are very small, the grand coalition could probably be formed, but in this case it makes no sense to improve the spectrum efficiency. So the grand coalition would never form.

Generally, the solving process of coalition formation game is too complicated, and not applicable in practice. Confronted with this problem, we propose a distributed algorithm making the process took place in a low-complexity manner. In the algorithm, merge and split rule is used for forming or breaking coalitions, while Pareto Optimality is used to compare the collections of coalitions.

A collection of coalitions is defined as a set of mutually disjoint coalitions which is denoted as $L = \{L_1, L_2, ..., L_i\}$. The collection in this paper also the partition of X. Given another collection $\overline{L} = \{\overline{L}_1, \overline{L}_2, ..., \overline{L}_{\overline{i}}\}$, the utility of player r in coalition $L_i \in L, 1 \leq i \leq I$ and coalition $\overline{L}_{\overline{i}} \in \overline{L}, 1 \leq \overline{i} \leq \overline{I}$ are $u_r(L) = u_r(L_i)$ and $u_r(\overline{L}) = u_r(\overline{L}_{\overline{i}})$, respectively. For all of the user, when $u_r(L) \geq u_r(\overline{L})$ happens with at least one strict inequality, then we define L is preferred over $\overline{L}$ by the Pareto Optimality. And the relationship is denoted as $L \sqsupset \overline{L}$. In order to find the stable collection, merge and split rule will be used [19]. When disjoint coalitions $\{L_1, L_2, ..., L_G\}$ in one collection have $\bigcup_{g=1}^{G} L_g \sqsupset \{L_1, L_2, ..., L_G\}$, while the utilities of the rest coalitions remain the same, these coalitions merge into one coalition $\{\bigcup_{g=1}^{G} L_g\}$. Otherwise when one coalition $\{\bigcup_{g=1}^{G} L_g\}$ has $\{L_1, L_2, ..., L_G\} \sqsupset \bigcup_{g=1}^{G} L_g$, the coalition is split into several coalitions $\{L_1, L_2, ..., L_G\}$. When merge or split operation happens, new collection is formed.

Use these rules, the energy-efficient uplink resource sharing algorithm can be described as follows:

1: The set of players is denoted as X, which includes all the users. Some data-requesting users can form clusters. Each cluster and single data-requesting users is a coalition.
2: The uplink user coalitions form a small collection $L_{0,U}$, while the cluster coalitions form a small collection $L_{0,CL}$. And the rest coalitions form a small collection L_R. So the initial collection of coalitions is $L_0 = L_{0,U} \cup L_{0,CL} \cup L_R$. According to (12), using the formulas mentioned above to get the coalition value set of each coalition. Note that the users in $L_{0,CL}$ and L_R is treated as normal downlink users.
3: Repeat the merge operation until all the coalitions have made their local merge decisions, then the resulting collection $\tilde{L}$ is obtained.
4: The collection $\tilde{L}$ accepts some split operations until it converge to a final collection L.

Starting from the collection L_0, we can always obtain the final collection using the algorithm. Every time one uplink user attempts to merge with a coalition which contains a cluster, the value set of the merged coalition will be calculated. Compare the utilities before and after the merge using Pareto Optimality, we can determine whether the merge is successful or not. According to the result of the merge operation, the collection of coalitions is obtained, remain the same or be different. Based on the fact that the number of coalitions in $L_{0,U}$ and $L_{0,CL}$ is finite, the process of the algorithm will end after several operations, and the final collection is obtained.

4   Numerical Results

In this section, the proposed algorithm is verified through computer simulations. Inspired by [20,21], the values of simulation parameters are summarized in Table 1. For each simulation, the location of the uplink users are generated randomly within the cell. The data-requesting users are distributed in a small area of the cell, which is easy to form D2D clusters. The data-requesting users distributed somewhere else work the same with these users. The channel gains $H_{i,j}$ between the transmitter i and the receiver j is calculated as:

$$H_{i,j} = 10lg(-h_{i,j}/10)$$
$$h_{i,j} = 32.4 + 20log10(d) + 20log10(f) \tag{13}$$

where $h_{i,j}$ follows the free space transmission loss formula, d is the distance between the two nodes, and f denotes the transmission frequency. For simplicity, the power consumption of circuit blocks for the transmitters and the receivers are set the same value. For comparison, two cases are considered: the first case is that all the users are in cellular mode, with no uplink resource reused; in the second case, every D2D cluster reuse one cellular user's uplink resource at most, and the relationship is one-to-one optimal according to exhaustive searches.

Figure 2 shows the uplink resource reusing relationships. The small hollow circle represents the data-requesting users, and the cross represents the uplink

Table 1. Simulation parameters

Cell radius	500m
Maximum distance within D2D cluster	$50\,m$
Maximum transmit power of the uplink users	$10\,mw$
Maximum transmit power of the data-requesting users	$0.1\,mw$
Constant circuit power	$10^{-4}\,mw$
Noise variance (σ^2) for 1 MHz bandwidth	$-144\,dbm$
Minimum throughput of the uplink users	$3.46\,bits/s/Hz$
Minimum throughput of the data-requesting users	$4.39\,bits/s/Hz$
The ratio of the power of PA to the transmit power	1.5
The compensate function parameter μ	0.5

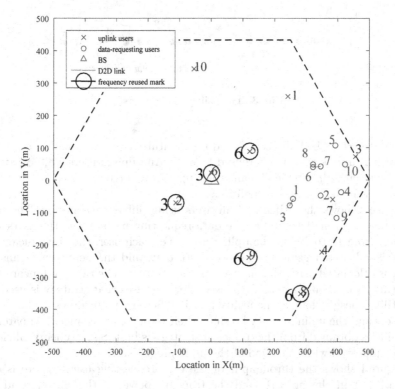

Fig. 2. D2D clusters reuse the cellular users uplink resource

user. The line between two data-requesting users means that they are in the same cluster, and the cluster head user is the one with the most lines. The cellular user besets with a big circle shares its uplink resource with a cluster, and the number near the big circle denotes the cluster head of the cluster. The data-requesting users 1 and 3 form a cluster called *clu*1, while data-requesting users 6, 7 and 8

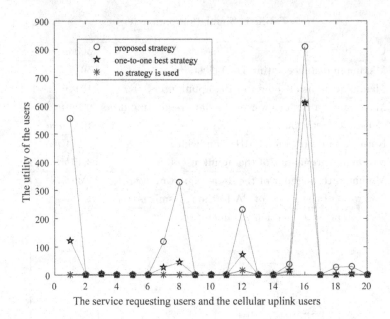

Fig. 3. The utility of the users

form cluster $clu2$. Uplink users 2 and 6 constitute a coalition with $clu1$, which means their uplink resource is reused by $clu1$. And uplink users 5,8,9 constitute a coalition with $clu2$, which means their uplink resource is reused by $clu2$. The rest users each constitutes a coalition.

Figure 3 shows the utilities of all users using different strategies. The users with numbers from 1 to 10 are the data-requesting users, while the users with numbers from 11 to 20 are the uplink users. For each user, the circle means the utility is calculated using the proposed strategy, and the pentagon means the utility is calculated using the one-to-one best strategy. The rice word means that the utility is 0 when no strategy is used. For the cases that strategy is used but the utility doesn't change, the utility is still 0. So only the utilities of the cluster tail users and the uplink users who share their resource are changed. Apparently, the utilities obtained from the proposed strategy is best. So the energy efficiency can be improved while the QoS of the users are satisfied.

Figure 4 shows the throughput of the data-requesting users. There is only one tail user in cluster $clu1$, and the transmit power of the cluster head and the uplink-sharing uplink users are designed basing on it, so the throughput of the tail user just meet the minimum requirement. However, there are two tail users in cluster $clu2$, when the most remote tail user 7 meet the minimum requirement, the throughput of the other tail user 8 would definitely be improved because of the short distance. The throughput marked with circle, pentagon, and rice word are respectively corresponded to the proposed strategy, the one-to-one

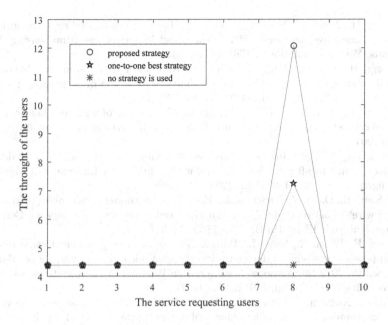

Fig. 4. The throughput of the users

best strategy and the situation with no cooperative strategy. As we can see, the proposed strategy shows the best performance of throughput.

From the simulations, the effectiveness of the game model and the algorithm is demonstrated.

5   Conclusion

In this paper, we proposed a new energy-efficient uplink resource sharing scheme. After establish and analyse the network model, we formulate the resource sharing problem as an NTU coalition formation game, and the algorithm with the merge and split rule is presented. The joint issue of uplink resource allocation and power management is solved. The simulation shows the scheme indeed improve the performance comparing with other methods.

Acknowledgment. This work is supported by the National Natural Science Foundation of China, No. 61271179, and the Beijing Municipal Science and Technology Commission research fund project "Research on 5G Network Architecture and Its Intelligent Management Technologies", No.D151100000115002.

References

1. Zhang, H., Jiang, C., Cheng, J., Leung, V.C.M.: Cooperative interference mitigation and handover management for heterogeneous cloud small cell networks. IEEE Wirel. Commun. **22**(3), 92–99 (2015)

2. Zhang, H., Jiang, C., Mao, X., Chen, H.H.: Interference-limited resource optimization in cognitive femtocells with fairness and imperfect spectrum sensing. IEEE Trans. Veh. Technol. **65**(3), 1761–1771 (2016)
3. Zhang, H., Liu, H., Jiang, C., Chu, X., Nallanathan, A., Wen, X.: A practical semidynamic clustering scheme using affinity propagation in cooperative picocells. IEEE Trans. Veh. Technol. **64**(9), 4372–4377 (2015)
4. Zhang, H., Chu, X., Guo, W., Wang, S.: Coexistence of wi-fi and heterogeneous small cell networks sharing unlicensed spectrum. IEEE Commun. Mag. **53**(3), 158–164 (2015)
5. Arvidsson, A., Manxing, D., Aurelius, A., Kihl, M.: Analysis of user demand patterns and locality for youtube traffic. In: 2013 25th International Teletraffic Congress (ITC), pp. 1–9. IEEE (2013)
6. Al-Kanj, L., Dawy, Z., Yaacoub, E.: Energy-aware cooperative content distribution over wireless networks: Design alternatives and implementation aspects. Commun. Surv. Tutorials IEEE **15**(4), 1736–1760 (2013)
7. Wang, F., Chen, X., Song, L., Han, Z., Zhang, B.: Energy-efficient radio resource and power allocation for device-to-device communication underlaying cellular networks. In: 2012 International Conference on Wireless Communications and Signal Processing (WCSP), pp. 1–6. IEEE (2012)
8. Qiu, X., Xuewen, L., Dong, K., Zhu, S.: Energy efficiency analysis in device-to-device communication underlaying cellular networks. In: 2013 IEEE Consumer Communications and Networking Conference (CCNC), pp. 625–630. IEEE (2013)
9. Mukherjee, A., Hottinen, A.: Energy-efficient device-to-device mimo underlay network with interference constraints. In: 2012 International ITG Workshop on Smart Antennas (WSA), pp. 105–109. IEEE (2012)
10. Wen, S., Zhu, X., Lin, Z., Zhang, X., Yang, D.: Energy efficient power allocation schemes for device-to-device (d2d) communication. In: 2013 IEEE 78th Vehicular Technology Conference (VTC Fall), pp. 1–5. IEEE (2013)
11. Wang, F., Chen, X., Song, L., Zhao, Q., Wang, X., Han, Z.: Energy-aware resource allocation for device-to-device underlay communication. In: 2013 IEEE International Conference on Communications (ICC), pp. 6076–6080. IEEE (2013)
12. Saad, W., Han, Z., Debbah, M., Hjorungnes, A., Basar, T.: Coalitional game theory for communication networks. Signal Process. Mag. IEEE **26**(5), 77–97 (2009)
13. Akkarajitsakul, K., Phunchongharn, P., Hossain, E., Bhargava, V.K.: Mode selection for energy-efficient d2d communications in lte-advanced networks: A coalitional game approach. In: 2012 IEEE International Conference on Communication Systems (ICCS), pp. 488–492. IEEE (2012)
14. Chia Hao, Y., Doppler, K., Ribeiro, C.B., Tirkkonen, O.: Resource sharing optimization for device-to-device communication underlaying cellular networks. Wirel. Commun. IEEE Trans. **10**(8), 2752–2763 (2011)
15. Wang, J., Zhu, D., Zhao, C., Li, J.C.F., Lei, M.: Resource sharing of underlaying device-to-device and uplink cellular communications. IEEE Commun. Lett. **17**(6), 1148–1151 (2013)
16. Belleschi, M., Fodor, G., Abrardo, A.: Performance analysis of a distributed resource allocation scheme for d2d communications. In: 2011 IEEE GLOBECOM Workshops (GC Wkshps), pp. 358–362 (2011)
17. Chien, C.P., Chen, Y.C., Hsieh, H.Y.: Exploiting spatial reuse gain through joint mode selection and resource allocation for underlay device-to-device communications. In: 2012 15th International Symposium on Wireless Personal Multimedia Communications (WPMC), pp. 80–84 (2012)

18. Dan, W., Wang, J., Rose, H., Cai, Y., Zhou, L.: Energy-efficient resource sharing for mobile device-to-device multimedia communications. Veh. Technol. IEEE Trans. **63**(5), 2093–2103 (2014)
19. Apt, K.R., Radzik, T.: Stable partitions in coalitional games. CoRR, abs/cs/0605132 (2006)
20. Wang, F., Song, L., Han, Z., Zhao, Q., Wang, X.: Joint scheduling and resource allocation for device-to-device underlay communication. In: 2013 IEEE Wireless Communications and Networking Conference (WCNC), pp. 134–139. IEEE (2013)
21. Chen, X., Song, L., Han, Z., Zhao, Q., Wang, X., Cheng, X., Jiao, B.: Efficiency resource allocation for device-to-device underlay communication systems: a reverse iterative combinatorial auction based approach. Sel. Areas Commun. IEEE J. **31**(9), 348–358 (2013)

Game Theory in Wireless Networks

Distributed Sharing of Base Stations for Greening: A Population Game Approach

Soohwan Lee$^{(\boxtimes)}$ and Yung Yi

Department of Electrical Engineering, KAIST, Daejeon, South Korea
soohwan@kaist.ac.kr, yiyung@kaist.edu

Abstract. Towards better QoSs and larger market share in highly competitive cellular network market, many mobile network operators (MNOs) aggressively invest in their base station (BS) deployment. As a result, BSs are densely deployed and incur a lot of energy consumptions, resulting in a large portion of operation cost. To save energy consumption, sharing BSs among different MNOs is a promising approach, where each user can be served from any BSs regardless of his or her original subscription, i.e., roaming. In this paper, we address the question of how many users should be roamed in a distributed manner with the goal of some sense of optimality. To answer this question, we take a population game approach, where we model flow-level dynamics of traffic and define an user association game among different MNOs. We prove that the game is an exact potential game with *'zero'-price-of-anarchy*. We develop a distributed algorithm that converges the NE (which is a socially optimal point) that can be used as a light-weight, dynamic user association algorithm.

Keywords: Greening · Base station sharing · Population game theory

1 Introduction

With increasing demands of mobile data traffic and large market competition among MNOs, most MNOs aggressively enhance their spectral efficiency by densely deploying BSs. As a result, the current BSs are densely deployed in many places, which incurs a lot of BS energy consumptions with large operating expenditures (OPEX). To save energy waste, BS sharing is a promising solution, where depending on the traffic conditions and the user locations, more energy-wise efficient association can be applied. However, without a suitable user association rule, the effects of BS sharing would not be impressive, which we address in this paper.

Main contribution. In this paper, we study user association policies under a certain roaming agreement among existing MNOs, where each MNO strategically tries to minimize their OPEX by regulating how users should behave for

This work was supported by Institute for Information & communications Technology Promotion (IITP) grant funded by the Korea government (MSIP) (B0126-15-1078).

© ICST Institute for Computer Sciences, Social Informatics and Telecommunications Engineering 2017
J. Cheng et al. (Eds.): GameNets 2016, LNICST 174, pp. 79–89, 2017.
DOI: 10.1007/978-3-319-47509-7_8

energy efficiency. The user association determines "how many users to roam?" and depends on some key factors such as roaming price and BS deployment of each MNO. The main contribution of this paper is summarized as follows:

- Determining user association in BS sharing is a very challenging problem and often losing tractability, when we consider user-level QoS. Thus, we take a population game approach, in which all users are categorized by different groups according to their adaptive modulation and coding (AMC) level and original subscription. This gives us a mathematically tractable framework with implications for a distributed user association algorithm. In our model, we take a flow-level performance (such as file-transfer delay) into account for measuring user-level QoS as done in [1].
- The challenges lying in analyzing the game are (a) the complex couplings among QoS, BS energy consumption and roaming fee, each of them depends on the other MNO's roaming price, heterogeneous BS deployment, and user distribution as well as (b) finding a distributed user association algorithm which has implementable complexity. We first show that user association population game is an exact potential game which has an NE with 'zero-price-of-anarchy'. Next, we also study the three evolutionary dynamics of game that provably converge to the NE, and propose a distributed user association algorithm inspired by the best response dynamic, which is one of the evolutionary dynamics considered. Finally, we verify the user-level QoS and greening effects in BS sharing through numerical simulations.

Related work. User association in BS sharing is proposed in [2–7]. Most work largely relies on packet-level throughput maximization [2,5,6] with an ideal assumption, in which all MNO have identically the same BS deployment for mathematical tractability. The authors in [3] only consider power consumption and roaming fee in BS sharing regardless of user-level QoS. The scope of the study in [4] is uplink BS sharing. Our work is mainly motivated by [7], the authors first consider a flow-level performance in BS sharing under some assumptions such as Shanon capacity based channel rate and existence of synchronous clock in user association. The main difference between [7] and our works is that we consider practical channel rate according to AMC-level with asynchronous user association clock. For the single MNO, the flow-level performance is considered in [1,8,9] and our work is motivated by [8–11] in the context of multiple MNOs. The authors in [9] take a population game for user association, in which all users behave to maximize one objective of the single MNOs, while our work considers competition among multiple MNOs with roaming fee in population game.

2   Model

2.1   System Model

Network and BS sharing service. We consider a set $\mathcal{M}$ of multiple MNOs, and each of them has operating BSs denoted by a set $\mathcal{B}_m$, respectively. For simple notations, we define a set of other MNOs $-m \doteq \mathcal{M} \setminus \{m\}$, and a set of entire BSs $\mathcal{B} \doteq \bigcup_{m \in \mathcal{M}} \mathcal{B}_m$, and we abuse the notation of m, where we use $m(b)$ to indicate the MNO who owned the BS b. For the service model, we consider that each user can subscribe only one MNO, but the user can be served from any BSs irrespective of her original subscription, if her MNO pays a certain roaming fee as done in [2–7]. Also, we assume that all MNOs do not differentiate in the service priority between roaming and unroaming users.

Users. We assume that there exist sufficiently many users in the cellular networks to consider the society of continuous mass of user groups called *classes*. The set of classes is denoted by $\mathcal{Q}$, and each class q has a mass denoted by d_q. We consider that each user in a class q commonly shares (i) original subscription, (ii) the set of supporting BSs, (iii) AMC-level from the supporting BSs, and (iv) traffic characteristics. Due to the need of denoting the set of classes that share original subscription, we occasionally use $\mathcal{Q}_m$, in which the original subscription of the classes $(q \in \mathcal{Q}_m)$ is the MNO m. Note that the subscription of user is mutually exclusive (i.e., $\mathcal{Q}_m \cap \mathcal{Q}_n = \phi$, for all $m \neq n$).

Traffic, capacity and loads. We assume that users in a class q have identically independent Poisson arrival traffic with rate λ_q, and its file size is independently distributed with mean $1/\mu_q$. Therefore, the unit mass in class q generates $\gamma_q = \lambda_q/\mu_q$ traffic intensity, and the class q totally generates $\gamma_q d_q$ traffic intensity. The users in q experience same data rate c_q^b when associating with BS b. Note that data rate only depends on AMC-level between the user and BS b, thus, user group would not be located on a point but on a region. For a pair of class q and BS b, we define *system-load intensity* as $\varrho_q^b \doteq \frac{\gamma_q}{c_q^b}$, which represents the service-time-portion of traffic intensity γ_q in BS b. For a given BS b, we introduce an association vector $\boldsymbol{y}^b \doteq (y_q^b : q \in \mathcal{Q})$, in which $y_q^b \in [0, d_q]$ denotes the fraction of class q's mass that are associated with BS b, where $\sum_{b \in \mathcal{B}} y_q^b = d_q$. For notational convenience later, we also use $\boldsymbol{y} \doteq (\boldsymbol{y}^b : b \in \mathcal{B})$ to denote the entire association vector. For a given association vector $\boldsymbol{y}$, we define *system-load* in BS b as $\rho^b(\boldsymbol{y}^b) \doteq \sum_{q \in \mathcal{Q}} \varrho_q^b y_q^b$.

3 Problem Formulation: Game

We consider a population game played by all users called *user association population game (UAPG)*, in which each user has an individual payoff function regulated by the MNO that he or she subscribes, and selfishly determines associating BS. Note that user association game implicitly reflects the selfish behavior of each MNO to minimize their cost (or equivalently maximize their revenue) by regulating the subscriber's payoff function under given roaming price $\boldsymbol{k} \doteq (k_m : m \in \mathcal{M})$ by a roaming agreement among MNOs a priori. In order to show the regulation rationale, we first describe a population game and we will compare the NE of the population game to that of conventional BS sharing game played by MNO (as done in [7]) in Sect. 4.

3.1 Social Objective

In order to include the selfish behavior of MNOs, we consider a social objective of UAPG as the potential function of BS sharing game as follows.

$$\mathcal{V}(\boldsymbol{y}) = -\sum_{b \in \mathcal{B}} \Big\{ \underbrace{\phi_\alpha(\boldsymbol{y}^b)}_{(a)} + \underbrace{\eta \mathcal{E}^b(\boldsymbol{y}^b)}_{(b)} + \underbrace{k_{m(b)} \sum_{q \notin \mathcal{Q}_{m(b)}} g^b(y_q^b)}_{(c)} \Big\}, \qquad (1)$$

where *(a)* is flow-level performance, *(b)* is BS power consumption, and *(c)* is roaming fee. In detail, *(a)* flow-level performance (such as file-transfer delay) is modeled by:

$$\phi_\alpha(\boldsymbol{y}^b) = \begin{cases} \frac{(1-\rho^b(\boldsymbol{y}^b))^{1-\alpha}}{\alpha-1}, & \text{if } \alpha \neq 1, \\ \log(\frac{1}{1-\rho^b(\boldsymbol{y}^b)}), & \text{if } \alpha = 1, \end{cases} \qquad (2)$$

where the parameter $\alpha \geq 0$ characterized cost for flow-level performance. It is well known that the function represents the summation of user rate when $\alpha = 0$, and the summation of average delay when $\alpha = 2$ by [1]. For BS power consumption *(b)*, we consider BS load proportional BS power consumption for each BS b, modeled by:

$$\mathcal{E}^b(\boldsymbol{y}^b) = \beta^b E^b \rho^b(\boldsymbol{y}^b) + (1 - \beta^b) E^b, \qquad (3)$$

where $\beta^b \in [0,1]$ is a parameter quantifying the portion of load proportional power and E^b is maximum BS power consumption when fully utilized (i.e., $\rho^b(\boldsymbol{y}^b) = 1$). Note that BS b is ideally *energy-proportional* when $\beta^b = 1$, but, β^b ranges from 0.5 to 0.8 in practical BSs [12]. In *(c)*, for a given BS b, the function $g^b(y_q^b)$ represents the summation of load and BS power consumption where the original subscription of class q is not the MNO who owns BS b (i.e., roaming traffic in class q) as follows.

$$g^b(y_q^b) = \varrho_q^b \cdot y_q^b + \eta \beta^b E^b \varrho_q^b \cdot y_q^b. \qquad (4)$$

Note that $\varrho_q^b \cdot y_q^b$ represents incurred load on BS b by the amount of y_q^b mass of class q. In (1), the parameter $\eta \geq 0$ trade off flow-level performance and BS power consumption. The large η implies MNOs give higher priority to BS power consumption than flow-level performance when operating cellular networks. The value k_m, which is given by some constant K, and is unit roaming price determined by each MNO when they make an agreement on roaming. Thus, social objective (1) represents negative total costs for entire traffic service (including roaming and unroaming) in whole cellular networks.

3.2   Payoff Function

We now introduce the payoff function for a class q in our population game as follows.

$$F_q^b(\boldsymbol{y}) \doteq \underbrace{\frac{-\varrho_q^b}{(1 - \rho^b(\boldsymbol{y}))^\alpha}}_{(i)} - \underbrace{\eta \beta^b E^b \varrho_q^b}_{(ii)} - \underbrace{k(b,q)\varrho_q^b(1 + \eta \beta^b E^b)}_{(iii)}, \qquad (5)$$

where $k(b,q)$ represents the unit roaming price of BS b's owner, when q is not the subscribers of the owner and 0 otherwise (i.e., if $q \notin \mathcal{Q}_{m(b)}$ then $k(b,q) = k_{m(b)}$, and $k(b,q) = 0$ for otherwise). The payoff function is composed of three part: (i) selfish QoS cost, (ii) BS power pricing, and (iii) roaming pricing.

(i) Selfish QoS cost: The first term describes selfish QoS cost motivated by flow-level performance cost as described in (2). Note that for $\alpha = 1$, this term represents to *conditional delay*, where the conditional delay is the expected file-transfer time that a user in class q experiences when she is associated with BS b as described in [9,13].

(ii) BS power pricing: The second term denotes the increments in BS power consumption when the unit mass of class q is associated with BS b. Note that for a user in class q, this term considered as a proportional factor of power increment when the user is associated with BS b, thus actual power increment is multiplication of user's mass x and this term (i.e., $\eta \beta^b E^b \varrho_q^b \cdot x$).

(iii) Roaming pricing: The third term corresponds to the incurred roaming fee by unit mass of class q. Similar to BS power pricing, a user in class q generates roaming fee according to her mass x with proportional to this term, if the user is associated with BS b (i.e., $k(b,q)\varrho_q^b(1 + \eta \beta^b E^b) \cdot x$).

4   Equilibrium Analysis

In this section, we analyze UAPG for which we exploit the potential function of the game. Primary issues that we are interested in include the existence of NE, price-of-anarchy, and the existence of a distributed user association algorithm which converges to the NE.

4.1   Price-of-Anarchy and Existence of Equilibrium

Prior to describe price-of-anarchy and equilibrium, we first show that our game is an exact potential game with a certain potential function that gives us the insight of price-of-anarchy and the existence of equilibrium.

Theorem 1. *The user association game is an exact potential game with the following potential function $V(\boldsymbol{y})$:*

$$V(\boldsymbol{y}) = -\sum_{b \in B} \left\{ \phi_\alpha(\boldsymbol{y}^b) + \eta \mathcal{E}^b(\boldsymbol{y}^b) + k_{m(b)} \sum_{q \notin \mathcal{Q}_m} g^b(y_q^b) \right\}. \qquad (6)$$

Proof. For a continuous player set (e.g., large population of player), it is suffice to show that there is a continuously differentiable function whose gradient for population is same as the payoff function of each class by [14]. The gradient of the potential function (6) for all population is given by:

$$\nabla_{\boldsymbol{y}} V(\boldsymbol{y}) = \left(\frac{\partial V(\boldsymbol{y})}{\partial y_q^b} : q \in \mathcal{Q}, b \in \mathcal{B} \right)$$

For all $q \in \mathcal{Q}$ and $b \in \mathcal{B}$, in the case $\alpha \neq 1$,

$$\frac{\partial V(\boldsymbol{y})}{\partial y_q^b} = -\frac{\partial}{\partial y_q^b} \Big[\sum_{b \in \mathcal{B}} \Big\{ \frac{(1 - \rho^b(\boldsymbol{y}^b))^{1-\alpha}}{\alpha - 1} + \eta(\beta^b E^b \rho^b(\boldsymbol{y}^b) + (1 - \beta^b) E^b)$$

$$+ k_{m(b)} \sum_{q \notin \mathcal{Q}_{m(b)}} \varrho_q^b y_q^b + \eta \beta^b E^b \varrho_q^b y_q^b \Big\} \Big]$$

$$= -\frac{\partial}{\partial y_q^b} \Big[\sum_{b \in \mathcal{B}} \Big\{ \frac{(1 - \sum_{q \in \mathcal{Q}} \varrho_q^b y_q^b)^{1-\alpha}}{\alpha - 1} + \eta(\beta^b E^b \sum_{q \in \mathcal{Q}} \varrho_q^b y_q^b + (1 - \beta^b) E^b)$$

$$+ k_{m(b)} \sum_{q \notin \mathcal{Q}_{m(b)}} \varrho_q^b y_q^b + \eta \beta^b E^b \varrho_q^b y_q^b \Big\} \Big]$$

$$= -\Big[-\varrho_q^b \Big(\frac{1-\alpha}{\alpha - 1} \Big) (1 - \sum_{q \in \mathcal{Q}} \varrho_q^b y_q^b)^{-\alpha} + \eta \beta^b E^b \varrho_q^b + k(b,q)(\varrho_q^b + \eta \beta^b E^b \varrho_q^b) \Big]$$

$$= -\varrho_q^b \Big[\frac{1}{(1 - \rho^b(\boldsymbol{y}))^{\alpha}} + \eta \beta^b E^b + k(b,q)(1 + \eta \beta^b E^b) \Big] = F_q^b(\boldsymbol{y}).$$

For the case $\alpha = 1$,

$$\frac{\partial V(\boldsymbol{y})}{\partial y_q^b} = -\frac{\partial}{\partial y_q^b} \Big[\sum_{b \in \mathcal{B}} \Big\{ \log \Big(\frac{1}{1 - \rho^b(\boldsymbol{y}^b)} \Big) + \eta(\beta^b E^b \rho^b(\boldsymbol{y}^b) + (1 - \beta^b) E^b)$$

$$+ k_{m(b)} \sum_{q \notin \mathcal{Q}_{m(b)}} \varrho_q^b y_q^b + \eta \beta^b E^b \varrho_q^b y_q^b \Big\} \Big]$$

$$= -\frac{\partial}{\partial y_q^b} \Big[\sum_{b \in \mathcal{B}} \Big\{ \log \Big(\frac{1}{1 - \sum_{q \in \mathcal{Q}} \varrho_q^b y_q^b} \Big) + \eta(\beta^b E^b \sum_{q \in \mathcal{Q}} \varrho_q^b y_q^b + (1 - \beta^b) E^b)$$

$$+ k_{m(b)} \sum_{q \notin \mathcal{Q}_{m(b)}} \varrho_q^b y_q^b + \eta \beta^b E^b \varrho_q^b y_q^b \Big\} \Big]$$

$$= -\Big[-\varrho_q^b \Big(\frac{-1}{(1 - \sum_{q \in \mathcal{Q}} \varrho_q^b y_q^b)^2} \Big) (1 - \sum_{q \in \mathcal{Q}} \varrho_q^b y_q^b) + \eta \beta^b E^b \varrho_q^b$$

$$+ k(b,q)(\varrho_q^b + \eta \beta^b E^b \varrho_q^b) \Big]$$

$$= -\varrho_q^b \Big[\frac{1}{(1 - \rho^b(\boldsymbol{y}))} + \eta \beta^b E^b + k(b,q)(1 + \eta \beta^b E^b) \Big] = F_q^b(\boldsymbol{y}),$$

which completes the proof.

Lemma 1. *The potential $V(\boldsymbol{y})$ is a concave function in $\boldsymbol{y}$.*

Proof. The functions, $V(\boldsymbol{y})$, $\phi_\alpha(\boldsymbol{y}^b)$, and $\mathcal{E}^b(\boldsymbol{y}^b)$ are convex functions in $\rho^b(\boldsymbol{y}^b)$, respectively, and $\rho^b(\boldsymbol{y}^b)$ is a weighted (ϱ_q^b) linear combination of $\boldsymbol{y}^b$. Thus, $V(\boldsymbol{y})$ and $\phi_\alpha(\boldsymbol{y})$ become convex functions in $\boldsymbol{y}$ by convex-preserving operation. The function $g^b(y_q^b)$ is definitely a convex function in y_q^b as described in (4). Thus, $V(\boldsymbol{y})$ is a concave function in $\boldsymbol{y}$ (by inversed sign) due to the property of convex preserving on summation.

Theorem 2. *User association game has an NE which has zero price-of-anarchy.*

Proof. The association vector $\boldsymbol{y}$ is bounded by the mass of each class q (i.e., $d_q \in [0, d_q]$). Thus, there is a global maximal point on the range of association vector, and the point is an NE by well know property of potential game [14], in which the NE should satisfy KKT conditions for a maximizer of the potential function $V(\boldsymbol{y})$. Zero price-of-anarchy is also easily verified by the potential function. Since the potential function (6) is exactly equal to the social objective as described in (1) and KKT conditions are necessary and sufficient condition for a global maximizer in concave function, the NE satisfying KKT conditions should be a global maximizer of the social objective.

Note that there could be multiple NEs in UAPG, because the NE only implies an assigned amount of population for all pairs of user groups and BSs, and the assigned population would be achieved by various user associations when we consider identical users who share the traffic characteristics and AMC-level in each class.

Rationality for MNOs. As we mentioned earlier, for all classes in MNO m (i.e., $q \in \mathcal{Q}_m$), m regulates the class q's payoff function to maximize their economical revenue (or minimize cost) in our game, while the MNO m hopefully behaves like the game, directly played by MNOs as done in [7]. Note that our game and the game played by MNOs have the exactly same potential function (i.e., equivalent game) for an arbitrary unit roaming price. Thus, the payoff function (5) is rational to each MNO, and implicitly considers the selfish behaviors of all MNOs for maximizing their revenue.

4.2   Evolutionary Dynamics and Distributed Association Algorithm

Developing a distributed algorithm for user association is important in practice, because, if it exists, high energy-efficiency can be achieved with low-cost operations of the networks. In this subsection, we propose a distributed user association algorithm, motivated by an *evolutionary dynamic* that converges to the NE in population game. For convenience in understanding, we first introduce three well-known evolutionary dynamics [15], Replicator dynamic, Brown-von Neumann-Nash (BNN) dynamic, and best response dynamic with the definitions and the convergence properties, and then, propose a distributed user algorithm that can work practical cellular networks.

Replicator dynamic. One of the best known dynamic in evolutionary game is the replicator dynamic, and its definition is as follows.

$$y_q^{b,t+1} = T_q^b(\boldsymbol{y}^t) \doteq y_q^{b,t}\left(F_q^b(\boldsymbol{y}^t) - \frac{1}{d_q}\sum_{b\in\mathcal{B}} y_q^{b,t} F_q^b(\boldsymbol{y}^t)\right), \tag{7}$$

where $\boldsymbol{y}^t$ is the social-state $\boldsymbol{y}$ at time step t, and the term $(F_q^b(\boldsymbol{y}^t) - \frac{1}{d_q}\sum_{b\in\mathcal{B}} y_q^{b,t} F_q^b(\boldsymbol{y}^t))$ is the excess payoff of strategy b in class q. Under replicator dynamic, a user randomly selects an opponent in the same class and changes her strategy to the strategy of opponent, if the payoff of the opponent strategy is higher than her own with a probability proportional to the payoff difference.

BNN dynamic. The definition of BNN dynamic is as follows.

$$y_q^{b,t+1} = T_q^b(\boldsymbol{y}^t) \doteq d_q k_q^b(\boldsymbol{y}^t) - y_q^b\sum_{b\in\mathcal{B}} k_q^b(\boldsymbol{y}^t), \tag{8}$$

where $k_q^b(\boldsymbol{y}^t) = \max\{F_q^b(\boldsymbol{y}^t) - \frac{1}{d_q}\sum_{b\in\mathcal{B}} y_q^{b,t} F_q^b(\boldsymbol{y}^t), 0\}$. In BNN dynamic, each user randomly chooses a strategy i and changes her strategy to i with a probability proportional to strategy i's excess payoff, if the payoff of i exceeds the payoff of her own at every updating strategy epoch.

Best response dynamic. In best response dynamic, each user selects her strategy that maximizes her payoff function for a given social-state $\boldsymbol{y}$ as follows.

$$y_q^{b,t+1} = T_q^b(\boldsymbol{y}^t) \doteq \arg\max_{b\in\mathcal{B}} F_q^b(\boldsymbol{y}^t) \tag{9}$$

Note that a user selects exact one pure strategy in best response dynamic, however, when we consider a class, in which infinitesimal users individually select their strategies, best response (9) behaves like a mixed strategy in population game.

Convergence. It is well-known by [14], a dynamic, satisfying *positive correlation (PC)* and *noncomplacency (NC)* conditions, converges to the NE in potential game which has a smooth potential function. The first condition, PC, states that payoff and drift rate of strategy in dynamic are positively correlated (i.e., weak-monotonicity in dynamic). The details of PC is $T(\boldsymbol{y}) \cdot F(\boldsymbol{y}) \doteq \sum_{q\in\mathcal{Q}}\sum_{b\in\mathcal{B}} T_q^b(\boldsymbol{y})F_q^b(\boldsymbol{y}) > 0$, whenever $V(\boldsymbol{y}) \neq 0$. For every trajectory of dynamic, the condition PC implies that (i) the potential function is weakly-increasing (i.e., $\frac{d}{dt}V(\boldsymbol{y}^t) = \nabla_{\boldsymbol{y}^t}V(\boldsymbol{y}^t) \cdot \boldsymbol{y}^t = T(\boldsymbol{y}^t) \cdot F(\boldsymbol{y}^t) \geq 0$), (ii) there is zero-drift for a stationary point (i.e. $T(\boldsymbol{y}^t) = 0$ whenever $\frac{d}{dt}V(\boldsymbol{y}^t) = 0$). Thus, all trajectories satisfying PC provably converge to a stationary point. However, all stationary points would not be NEs, where the points are either local maximizer or boundary of potential function. The condition, NC, guarantees that a stationary point should be a NE of potential game. By the studies in [14,15], it is verified that BNN and best response dynamic satisfy both PC and NC, but replicator dynamic only satisfies PC in the potential game. For more detail, we refer the readers to [14,15].

Distributed user association algorithm. Low signaling overheads is important in practice. In UAPG, the best response dynamic seems to require less information than the others. In detail, best response dynamic only requires social state y, but the others require additional information such as average payoff and opponent's payoff. Thus, we propose a distributed algorithm motivated by best response dynamic.

Distributed user association algorithm

BS algorithm. For every changes in user association, each BS b updates $\rho^{b,t}$ as follows.

$$\rho^{b,t} = \sum_{q \in \mathcal{Q}} \varrho_q^b \cdot y_q^b, \tag{10}$$

and exchange $(\rho^{b,t}, k_{m(b)})$ to all BS in the neighboring BS set, denoted by $\mathcal{N}(b)$. After exchanging the information, each BS broadcasts $\rho^{\mathcal{N}(b),t}, \rho^{b,t}$ and k to all (associated) users.

User algorithm. For a user in some class q, at every *association clock[a]* ticking, the user associates with a BS that satisfies following:

$$\underset{b \in \{i\} \cup \mathcal{N}(i)}{\arg\max} \; -\varrho_q^i \left\{ \frac{1}{(1 - \rho^{i,t})^\alpha} + \eta \beta^i E^i + k(i,q)(1 + \eta \beta^i E^i) \right\}, \tag{11}$$

---

[a] We consider each user has an individual clock for determining user association. In detail, this clock would be implemented by many ways such as Poisson clock, and flow arrival and departure time.

Theorem 3. *The distributed user association algorithm converges to the NE.*

Proof. Our algorithm is a practical version of best response dynamic which satisfies both (i) PC and (ii) NC.

5   Numerical Analysis

In this section, we verify the greening effects of our algorithm inspired by the analysis in the population game framework. In all simulations, we consider a duopoly market (i.e., 2 MNOs denoted by m and n) in 0.5 Km by 0.5 km square area, in which MNO m and n have 1 BS denoted by BS1 and BS2, respectively, where BS1 and BS2 are located at $(0,0)$ and $(0.5,0.5)$[1], respectively. We consider that users are uniformly distributed in the square area while generating homogeneous traffic requests. For data rate c_q^b, we refer to the pairs of data rate

---

[1] The unit of axis is km.

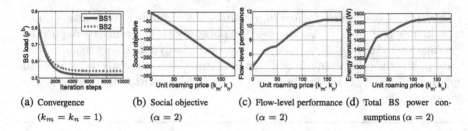

(a) Convergence
($k_m = k_n = 1$)

(b) Social objective
($\alpha = 2$)

(c) Flow-level performance
($\alpha = 2$)

(d) Total BS power consumptions ($\alpha = 2$)

Fig. 1. Various results of BS sharing

and AMC-level in Mobile WiMAX [16]. We consider the case when all MNOs adopt a same unit roaming price (i.e., $k_m = k_n$) for roamed traffic due to the symmetry property in unit roaming price under symmetric BS deployment and identical user characteristics.

We first verify the convergence of our algorithm (see Fig. 1(a)). In the figure, the initial points are the BS loads of conventional non-BS sharing and each BS load rapidly decays from the initial point until it converges with iterations. In Figs. 1(b), (c) and (d), we show the impact of our algorithm in terms of potential function, flow-level performance, and BS power consumptions according to given unit roaming price k. The result in Fig. 1(b) shows that social objective is maximized when each MNO assigns zero-unit roaming price and it decreases as k increases due to the raised roaming price. As shown in Figs. 1(c) and (d), flow-level performance (i.e., delay when $\alpha = 2$) and total BS power consumptions are increased by expensive roaming price, and finally converge to that in conventional non-BS sharing, since no one is interested in roaming when highly expensive roaming price (e.g., $k_m = k_n \geq 150$) is applied.

6   Conclusions

In this paper, we studied BS sharing under a fixed roaming price using a population game-theoretic approach, and we proposed a practical user association algorithm motivated by an evolutionary dynamic, which is best response dynamic. We further demonstrated that a significant amount of delay and of energy consumption would be reduced by the proposed algorithm.

References

1. Kim, H., De Veciana, G., Yang, X., Venkatachalam, M.: Distributed α-optimal user association and cell load balancing in wireless networks. IEEE/ACM Trans. Networking **20**(1), 177–190 (2012)
2. Marsan, M.A., Meo, M.: Energy efficient management of two cellular access networks. Sigmetrics Perform. Eval. Rev. **37**(4), 69–73 (2010)
3. Yanan, B., Jian, W., Sheng, Z., Zhisheng, N.: Bayesian mechanism based inter-operator base station sharing for energy saving. In: Proceedings of IEEE ICC (2015)

4. Wu, Y., Zhu, Q., Huang, J., Tsang, D.: Revenue sharing based resource allocation for dynamic spectrum access networks. IEEE J. Sel. Areas Commun. **32**(11), 2280–2296 (2014)
5. Leng, B., Mansourifard, P., Krishnamachari, B.: Microeconomic analysis of base-station sharing in green cellular networks. In: Proceedings of IEEE Infocom (2014)
6. Bousia, A., Kartsakli, E., Antonopoulos, A., Alonso, L., Verikoukis, C.: Game theoretic infrastructure sharing in multi-operator cellular networks. IEEE Trans. Veh. Technol. **99**, 1 (2015)
7. Lee, S., Moon, S., Yi, Y.: On greening cellular networks by sharing base stations: a game-theoretic approach. In: Proceedings of Valuetools (2015)
8. Son, K., Kim, H., Yi, Y., Krishnamachari, B.: Base station operation and user association mechanisms for energy-delay tradeoffs in green cellular networks. IEEE J. Sel. Areas Commun. **29**(8), 1525–1536 (2011)
9. Moon, S., Kim, H., Yi, Y.: Brute: Energy-efficient user association in cellular networks from population game perspective. IEEE Trans. Wirel. Commun. **99**, 1 (2015)
10. Kwak, J., Son, K., Yi, Y., Chong, S.: Greening effect of spatio-temporal power sharing policies in cellular networks with energy constraints. IEEE Trans. Wirel. Commun. **11**(12), 4405–4415 (2012)
11. Lee, S., Son, K., Gong, H., Yi, Y.: Base station association in wireless cellular networks: An emulation based approach. IEEE Trans. Wirel. Commun. **11**(8), 2720–2729 (2012)
12. Peng, C., Lee, S.-B., Lu, S., Luo, H., Li, H.: Traffic-driven power saving in operational 3g cellular networks. In: Proceedings of ACM MobiCom (2011)
13. Altman, E., Ayesta, U., Prabhu, B.: Load balancing in processor sharing systems. Telecommun. Syst. **47**(1–2), 35–48 (2011)
14. Sandholm, W.H.: Potential games with continuous player sets. J. Econ. Theor. **97**(1), 81–108 (2001)
15. Sandholm, W.H.: Population games and evolutionary dynamics. MIT press, Cambridge (2010)
16. Mobile WiMAX - part i a technical overview and performance evaluation bibtex. WiMAX Forum (2006)

Weighted Voting Game Based Relay Node Managemnet in VANETs

Elham Dehghan Biyar[✉] and Berk Canberk

Department of Computer Engineering, Istanbul Technical University,
Ayazaga Campus, 34469 Istanbul, Turkey
{dehghanbiyar,canberk}@itu.edu.tr

Abstract. In traditional Vehicular Ad Hoc Networks (VANETs) deployments, permanent and robust connection establishment to road side units (RSU) has arisen as a crucial problem. Here it is a known fact that, this challenge has been triggered by high mobility pattern of vehicles. To handle this problem, optimal relay vehicle selection can be seen as an efficient solution. To this end, in this paper, we propose a novel optimal and fair relay vehicle selection algorithm based on weighted voting game. In our game theoretic approach, relay vehicle selections have been performed by various cooperative coalitions. Note that game theory is a perfect tool while designing such an algorithm as it is a formal applied mathematical tool to analyze and model complicated situations of interactive decision making. Our proposed weighted voting game algorithm can achieve fair and optimal results as well as increasing throughput and decreasing message transmission delay during packet dissemination as a result of using Banzhaf power measure. Performance evaluation results depicted that compared to non-cooperative methods, throughput increases by 24.4% and message dissemination delay decreases by 18%.

Keywords: Game Theory · Pay-off function · VANETs · Weighted Voting Game · Banzhaf power measure · Fair relay

1 Introduction

With the recent advances in Information and Communications Technology (ICT), VANET has become an important concept in order to provide efficient and convenient road trips for drivers by obtaining required information along the road. This information can vary from infotainment to traffic efficiency, on demand application management and updating information [1,2]. For obtaining all the mentioned infromation, packet dissemination all along the road is important. Due to the mobile nature of vehicles that results in dynamic topological changes, establishing permanent and robust connection with road side units for maintaining connectivity has arisen as a challenging issue. Furthermore, transmission failure of actual amount of packets during data transfer can happen because of Doppler effect, loss of signal, dissimilar speed of send and receive,

© ICST Institute for Computer Sciences, Social Informatics and Telecommunications Engineering 2017
J. Cheng et al. (Eds.): GameNets 2016, LNICST 174, pp. 90–100, 2017.
DOI: 10.1007/978-3-319-47509-7_9

and bandwidth limitation of RSU [3]. To overcome these problems and for efficient management of optimal connectivity, it is essential to use some of vehicles as a relay for compensation and reinforcement of connection. While selecting the optimal relay vehicle, we need to take into account several factors such as quality of service (QoS) [4,5], system performance, cost-efficiency role of chosen node and fairnes [6,7]. Providing a satisfying method for this selection is required and it is a challenging issue due to dynamic environment of roads.

There are works which have dealt with relay node selection. In [8] source nodes try to find the most appropriate relay node based on self optimizing algorithm called SLA. In [9] distributed relay node selection provides responsibility of rebroadcasting of alert messages to further distances. They have used bidirectional stable communication algorithm for selecting set of qualified relay nodes. The author has focused on quality of relay nodes and has not considered other metrics and optimization methods while selecting the optimal relay among multiple options. Besides, none of them did a work using Game Theory. In [3] coalitional game theory approach for solving cost-efficient content downloading has been proposed. In [10] relay vehicle selection based on game theory is proposed, pay-off functions are designed with respect to some metrics, and an optimal matching problem has been solved using Kuhn-Munkres algorithm. Although game theory has been used to optimal relay vehicle selection, defining Nash equilibrium point for cooperation is missing. Moreover, fairness has not been considered. In this paper, an algorithm based on Weighted Voting Game will be introduced that we believe is more efficient for optimization and fairness. All aforementioned works, have mainly focused on quality of chosen relay node, but few works are focused on choosing a relay among multiple number of eligible relays. Moreover, in these proposed methods fair relay node selection haven't been considered. One of the notable issues in network management is fairness [6]. In this paper, fairness is defined as an impartial relay assignment in a way that it also optimizes pay-off functions for various individual anchor users. In addition, we believe that Game Theory should be used in relay vehicle selection because it is a formal applied mathematical tool to analyze and model complicated situations of interactive decision making [11]. There are several decision makers with various intentions, which decision of each one effects the overall result of decision making process [11]. In this paper Weighted Voting Game is proposed, which is a popular model of interactive decision making in cooperative games. This model has recently been used in a wide range of research areas such as economy, science, and management [12,13]. Besides, Banzhaf power measure to designate both fairness and optimization parameters for relay vehicle selection has been introduced. Moreover, in each game with bounded number of players there exist at least one Nash equilibrium point [14,15], so that existence of Nash equilibrium point is demonstrated through using this algorithm. The main Contributions of this paper, are as follows:

- We propose an algorithm based on Weighted Voting Game to choose fair and optimal relay vehicle for anchor vehicles.

- We show that after using Weighted Voting Game (WVG) algorithm, obtained solution is a Nash Equilibrium (NE) point.
- We calculate pay off functions for both anchor and relay vehicles.
- We introduce a method to choose a fair relay node to acquire fairness.

2   Network Architecture

The abstract network is shown in Fig. 1 that consists of RSU, N anchor vehicle and M relay vehicle. These steps will occur respectively:

1. We have a message to give from RSU to vehicles.
2. If the vehicle is in the coverage area, it will try to get that message directly from RSU. In the case of failure for direct connection, it will try to use relay nodes.
3. For the vehicles which are not in the coverage area of RSU, it is not possible to get this message unless they enter that area.
4. With the proposed algorithm vehicles disseminate the RSU message among all other vehicles, even if they are not in the coverage area. After all we propose an algorithm to select the best relay node to get the message and it will be based on weighted voting game algorithm.

3   System Model

As shown in Fig. 2, we have 2 number of anchor vehicles (D, E) and 3 number of relay vehicles (A, B, C). Anchor vehicles are the vehicles that fail to establish connection directly to road side units so they try to find optimal relay node for packet dissemination. In our scenario relay nodes are the nodes that help anchors to preserve connectivity. While (D, E) enter coverage area of RSU, they try to connect to RSU in the case of connection setup failure, as shown in Fig. 2(b) they use one of (A, B, C) nodes as a relay to establish connection. One of these optimal nodes will be chosen by our proposed algorithm. Another scenario is depicted in Fig. 3 where vehicles (F, G) from other lane join to the main lane and want to access to RSU's information. In this phase (F, G) can form coalition with (A, B, C) which have RSU header as their origin [16]. By our proposed algorithm one of (A, B, C) nodes can be choosen as an optimal relay for the anchors.

Each game is consist of players, action profiles, preferences and pay-off functions. In this paper multiple anchor vehicles and multiple relay vehicles are the players of the weighted voting game and will be denoted by M and N respectively. Overall we need to consider two kind of relays:

1. Local relay node: In the presence of RSU, a relay can be selected, when vehicles fail to connect directly to RSU.
2. Mobile relay nodes: After exiting from RSU coverage area, relay node can be selected for connection maintenance. Furthermore, relays with appended RSU header [16] can act as a small RSU for other vehicles.

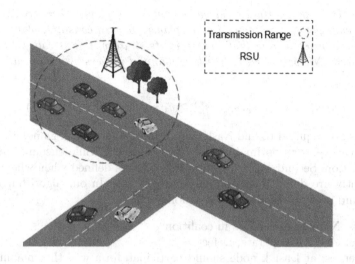

Fig. 1. Network topology

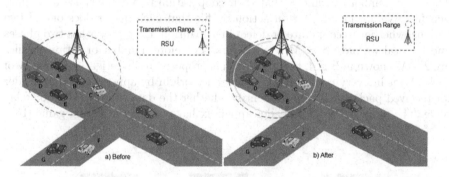

Fig. 2. Network topology example

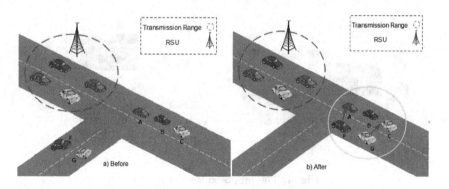

Fig. 3. Network topology example

Definition 1: In the normal-form game that has n-player $G = \{S_1, ..., S_n\}$, if player S_i changes its strategy and other players remain constant, player S_i can't acquire more benefits from that change in its strategy, this is called Nash equilibrium point. NE point will tell us how many of players are cooperating in the coalition [17].

$$u_i(s_1^*, ..., s_{i-1}^*, s_i^*, s_{i+1}^*, s_n^*) \geq u_i(s_1^*, ..., s_{i-1}^*, s_i, s_{i+1}^*, s_n^*) \tag{1}$$

Two steps are required to find Nash equilibrium point in each game. First, dealing with each players optimal strategy consecutively, while encountering other player's action. Second, a Nash equilibrium point is defined when whole players concurrently are doing their optimal approach [15]. In our algorithm Players, Actions and preferences are defined as follows:

1. Players: Number of vehicles in coalition
2. Actions: A = { Cooperate, Defect }
3. Preferences: at least k node should participate in a way that maximize the total utility, so in this case we will have NE point.

There are number of vehicles that their cooperation in a game cause to acquire benefits. We are looking for an action profile that each player does one of two actions, whether to cooperate or defect in a coalition. Each player has two phases and in total we have $S = M + N$ players. Therefore, number of action profiles are 2^S. As shown in Fig. 4, this algorithm is proposed to investigate optimality of each player in every coalition. Detecting data origin by appending small header for received packet will help to identify whether the data origin is RSU or relay node. The nodes with RSU header appendix have priority in voting game [16].

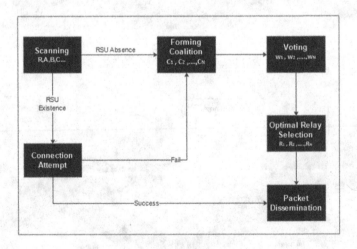

Fig. 4. The proposed system model

3.1 Scanning

In order to find neighbors for packet dissemination, utility functions for each present node must be calculated to decide whether to setup direct connection with RSU or looking for relay node. Defection in coalition for direct connection with RSU must be evaluated at first. Regarding to the amount of pay-off function, as represented as U in Eq. (2) defection may cause connection establishment or failure with RSU [17].

$$U = R - C \tag{2}$$

where R represents revenue of relay or anchor vehicle through using RSU or relays and C shows data forwarding expenses. In this step, the amount of R^r for relay vehicle can be expressed as:

$$R^r = P_{jRSU} B_j F_{ij} \tag{3}$$

where P is successful transmission probability of links between jth relay vehicle and RSU, B is bandwidth availability of RSU and F is bandwidth characteristic of relay vehicle [10].

After connection establishment with roadside unit these connected nodes can act as relay nodes. Afterward setting data origin to RSU should happen [16]. In the case of connection establishment failure with the RSU, Packet dissemination through RSU will fail.

3.2 Connection Attempt

In this step, as depicted in Fig. 4, some of vehicles start packet dissemination successfully. Otherwise, other steps are taking place.

3.3 Forming Coalition

Definition 2: Coalition is the subset of players that all vote in the same way. The number of all possible coalitions are; $2^N - 1$ [18].

By neighbor detection, anchor nodes can form coalition with relay node candidates, which are calculated and identified in previous steps.

3.4 Voting

Definition 3: Weighted voting game voters are unequal in the number of votes they control, it is depicted as:

$$[q : w_1, w_2, \ldots, w_n]$$

where q is quota and w is weight [18].

Definition 4: Winning coalitions are the coalitions that have enough number of votes to win. Voters are unequal in the number of votes they control [18].

In voting step, pay-off function computation for every neighbor that holds RSU header, is required. For this aim amount of R^a can be expressed as:

$$R^a = \frac{P_{ij}B_iAD}{T} \tag{4}$$

where P is Successful transmission probability of links between ith Relay Vehicle and jth anchor node, B is bandwidth availability of ith relay, T is data transmission time, D is distance between relay and anchor node, A is attainable rate of the link between ith relay and jth anchor [10]. The cost of service by each relay and RSU or anchor, can be modeled as a function of unit price denoted by α and spended bandwith resource [10]:

$$C = \alpha_i F_{ij} \tag{5}$$

After calculating pay-off functions, one of anchor nodes and whole relays will be chosen to decide upon the optimal relay for specific anchor. In addition, utility functions have been assigned as weights of relays. In each winning coalition the range of the quota is defined as [18]:

$$\frac{w_1 + w_2 + \ldots + w_n}{2} \leq q \leq w_1 + w_2 + \ldots + w_n \tag{6}$$

3.5   Optimal Relay Selection

In this step after calculating pay-off functions for anchor vehicles:

$$U^a = R^a - C \tag{7}$$

And assigning this pay-off function as a weight of our voting game

$$w = U^a \tag{8}$$

quota will be set to

$$q = \frac{w_1 + w_2 + \ldots + w_n}{2} \tag{9}$$

In our algorithm the amount of q is:

$$q = \frac{u_1^a + u_2^a + \ldots + u_n^a}{2} \tag{10}$$

Furthermore to acquire fairness while applying weights, Banzhaf power measure has been introduced.

Definition 5: Critical player is the player that eliminating it's weight from the whole votes cause the coalition turns into loosing one and the number of remaining votes fail to pass the quota. Some voters are more powerful [18].

Required steps for Banzhaf power measure calculation are listed as follows [18,19]:

1. Listing all achievable wining coalitions.
2. Determining critical players.
3. In succession check the number of times players are critical, this amount is shown with B_i notation
4. Calculate total number of times that players are critical $\sum_{i=1}^{N} B_i$.
5. The proportion of $\beta = \frac{B_i}{\sum_{i=1}^{N} B_i}$ gives Banzhaf power index.

The most powerful node with higher β is set to be the relay of coalition. These steps will be repeated for all other anchors. For choosing relay node in the absence of RSUs or direct connection failure with RSU, last two steps which are voting and optimal relay selection will be repeated. After these procedures each node is doing its own best strategy. After all, it is proved that the outcome of the offered algorithm is a Nash equilibrium point.

3.6 Packet Dissemination

In packet dissemination, after passing all previous steps and choosing fair and optimal relay, packet dissemination among all vehicles will start. Choosing optimal relay vehicle is important, consequently anchor vehicle's pay-off function is related to throughput function. Considering revenue function of anchor vehicle which is calculated in Eq. (4). Throughput is the rate of received packets at the destination over communication channel [20]. Our objective is to maximize throughput and minimizing message transmission delay respectively. As shown below, throughput is a function of:

$$T(u^a) = \frac{P_{ij}B_i AD}{T} - \alpha_i F_{ij} \tag{11}$$

where T is throughput.

4 Performance Evaluation

The performance of our proposed cooperative weighted voting game algorithm will be evaluated and compared with non-cooperative approach by using Matlab. In our simulation a road of 5 Km that has allocated road side units and number of vehicles that varies from 20 to 90 with randomly distributed velocities has been considered. The simulation parameters are all shown in Table 1. The aim of this algorithm is choosing a fair and efficient relay for anchor vehicles. The simulation results have validated our analysis and demonstrate better throughput and transmission delay outcomes. Our proposed algorithm can achieve 24.4% of increment in throughput as well as 18% reduction in transmission delay compared to non-cooperative approach.

Figure 5, demonstrates total throughput of all vehicles versus number of vehicles. It can be observed that during relay vehicle selection, by using Banzhaf power measure, pay-off function increases. Fair relay will be chosen considering

Table 1. Parameters for simulation

Number of vehicles	$[20, 90]$
Road length	$5\,km$
Number of lanes	2
RSU coverage area diameter	$60 \sim 350\,m$
Max speed	$25\,m/s$
Min speed	$45\,m/s$
Pricing factor α	100
SNR of transmitter	$10\,db$
Number of simulation iteration	100

Eq. 11. Pay-off function increment results in better throughput. To be more specific using weighted voting game algorithm causes optimal relay vehicle selection, which also has maximum available bandwidth B and successful transmission probability P.

Figure 6, displays average transmission delay versus number of vehicles. It is noted that by increasing number of vehicles, more coalition will occur, besides probability of successful reception is the other parameters that has been considered within our proposed algorithm which causes to consume more time in our voting game algorithm to find optimal relay. However, as depicted, this incremental results are less than non-cooperative approach.

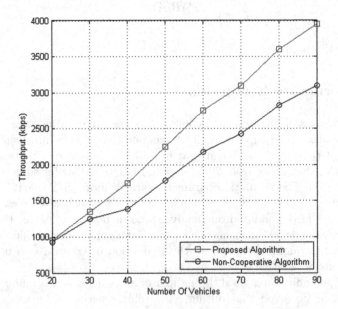

Fig. 5. Total throughput

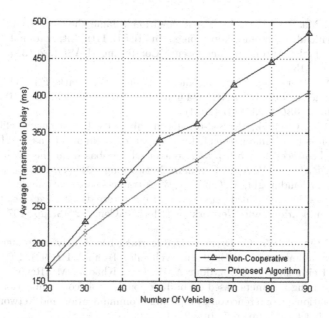

Fig. 6. Average transmission delay

5   Conclusion

A game theory based relay vehicle selection algorithm, based on weighted voting game, by applying Banzhaf power measure has been introduced. The proposed algorithm selects optimal and fair relay vehicle for packet dissemination by using pay-off functions that are derived for both anchor and relay vehicles. Our proposed algorithm, as shown in Fig. 4, is consists of scanning, connection attempt, forming coalition, voting, optimal relay selection and packet dissemination modules. Moreover, our proposed weighted voting game algorithm can achieve fair and optimal results, as well as, increasing throughput and decreasing message transmission delay during packet dissemination as a result of using Banzhaf power measure. Performance evaluation results illustrated that compared to non-cooperative methods, throughput increases by 24.4% and message dissemination delay decreases by 18%.

References

1. Karagiannis, G., Altintas, O., Ekici, E., Heijenk, G., Jarupan, B., Lin, K., Weil, T.: Vehicular networking: a survey and tutorial on requirements, architectures, challenges, standards and solutions. IEEE Commun. Surv. Tutorials **13**(4), 584–616 (2011)
2. Bozkaya, E., Erel, M., Canberk, B.: Connectivity provisioning using cognitive channel selection in vehicular networks. In: Mitton, N., Gallais, A., Kantarci, M.E., Papavassiliou, S. (eds.) Ad Hoc Networks, pp. 169–179. Springer, Heidelberg (2014)

3. Tong, L., Ma, L., Li, L., Li, M.: A coalitional game theoretical model for content downloading in multihop vanets. In: IEEE 11th International Conference on Dependable, Autonomic and Secure Computing (DASC), 2013, pp. 627–632, December 2013

4. Cheng, N., Zhang, N., Ning, L., Xuemin Shen, J.W., Mark, F.L.: Opportunistic spectrum access for CR-VANETs: a game-theoretic approach. IEEE Trans. Vehicular Technol. **63**(1), 237–251 (2014)

5. Akgül, Ö.U., Canberk, B.: Self-organized things (SoT): an energy efficient next generation network management. In: Computer Communications (2014)

6. Kim, S.: Femtocell network power control scheme based on the weighted voting game. EURASIP J. Wireless Commun. Networking **2013**(1), 1–9 (2013)

7. Arslan, Z., Alemdaroglu, A., Canberk, B.: A traffic-aware controller design for next generation software defined networks. In: First International Black Sea Conference on Communications and Networking (BlackSeaCom), 2013, pp. 167–171. IEEE (2013)

8. Chen, Z., Lin, T., Wu, C.: Decentralized learning-based relay assignment for cooperative communications. IEEE Trans. Vehicular Technol. **PP**(99), 1 (2015)

9. Hussain Rehman, O.M., Bourdoucen, H., Ould-Khaoua, M.: Relay selection for alert messaging in vanets based on bi-directional stable communication approach. In: International Conference on Computing, Communication and Networking Technologies (ICCCNT), pp. 1–7, July 2014

10. Rong Chai, L.V., Yuan, B.Y., Chen, Q.: Cooperative game based relay vehicle selection algorithm for vanets. In: 14th International Symposium on Communications and Information Technologies (ISCIT), pp. 30–34, September 2014

11. Maschler, M., Zamir, S., Solan, E.: Game Theory. Cambridge University Press, Cambridge (2013)

12. Aziz, H., Bachrach, Y., Elkind, E., Paterson, M.: False-name manipulations in weighted voting games. J. Artif. Intell. Res. **40**, 57–93 (2011)

13. Zuckerman, M., Faliszewski, P., Bachrach, Y., Elkind, E.: Manipulating the quota in weighted voting games. In: AAAI, vol. 8, pp. 215–220 (2008)

14. Jiang, A.X., Leyton-Brown, K.: A tutorial on the proof of the existence of nash equilibria. University of British Columbia Technical Report TR-2007-25. pdf (2009)

15. Benmammar, B., Krief, F.: Game theory applications in wireless networks: a survey. In: Proceeding 13th International Conference on Software Engineering, Parallel and Distributed Systems (SEPADS 2014), Gdansk, Poland, May, pp. 15–17 (2014)

16. Saad, W., Han, Z., Hjørungnes, A., Niyato, D., Hossain, E.: Coalition formation games for distributed cooperation among roadside units in vehicular networks. IEEE J. Selected Areas Commun. **29**(1), 48–60 (2011)

17. Gibbons, R.: A Primer in Game Theory. Harvester Wheatsheaf, Hertfordshire (1992)

18. Tannenbaum, P., Arnold, R.: Excursions in Modern Mathematics. Prentice Hall, Upper Saddle River (1998)

19. Laruelle, A., Valenciano, F.: Voting and Collective Decision-Making: Bargaining and Power. Cambridge Univ Press, Cambridge (2008)

20. Atayero, A.A.A.: Integrated Models for Information Communication Systems and Networks: Design and Development: Design and Development. IGI Global (2013)

The Study and Field Trial
of Coordinated Multi-point Techniques
in Heterogeneous Network

Yao Wei$^{(\boxtimes)}$, Shangkun Xiong, Qingyang Wang, and Ke Yin

Mobile Communication Research Department, Guangzhou Research Institute
of China Telecom Co., Ltd., Guangzhou 510630, China
wybby60@gmail.com, {xiongsk,wangqy,yink}@gsta.com

Abstract. With the development of mobile internet service, Long Term Evolution (LTE) system provides higher data rate and better user experience. One way to provide higher bitrates is to exploit or mitigate the interference by cooperation between sectors or different sites. Coordinated Multi-Point (CoMP) is one of the promising concepts to improve cell edge user data rate and spectral efficiency firstly introduced in LTE Release 11. In this paper, the principle and challenges of CoMP are introduced, and also the performance and results of CoMP based on field trial are given, especially the gain of cell edge user in heterogeneous network are analyzed. Finally, several proposals and suggestions of CoMP application are given in the end of this paper.

Keywords: Coordinated Multi-point · Joint reception · Coordinated scheduling

1 Introduction

LTE use MIMO-OFDM to achieve improved spectral efficiency within one cell [1–3]. With the evolution of LTE, new features are introduced in latest releases of the 3GPP specifications. One method coordination of eNBs to avoid interference and constructive exploitation of interference through coherent eNB cooperation is done. The cooperation techniques aim to avoid or exploit interference in order to improve the cell edge and average data rates. CoMP can be applied both in the uplink and downlink [4].

One of the fundamental differences between CoMP Multi-User (MU) MIMO systems and single-cell MU MIMO systems lies in the per base station power constraint [5]. By using CoMP, coherent transmission with coordinated base stations can significantly improve both the cell average throughput and the cell edge throughput. In CoMP a number of TX (transmit) points provide coordinated transmission in the DL, and a number of RX (receive) points provide coordinated reception in the UL. The set of TX/RX-points can either be at different locations, or co-sited but providing coverage in different sectors, they can also belong to the same or different eNBs [6].

CoMP is firstly introduced in 3GPP technical report 36.814 in February 2009, and officially compiled in Release 11 [7]. Rel.11 standardized uplink CoMP. The feature is transparent to UE, so it can also work in Rel.8 network. Rel.11 enhanced inter-cell

J. Cheng et al. (Eds.): GameNets 2016, LNICST 174, pp. 101–109, 2017.
DOI: 10.1007/978-3-319-47509-7_10

DMRS on PUSCH, PUCCH and SRS. For downlink CoMP, Rel.11 introduced the VCI (virtual cell identity), different VCIs can be configured in one cell, and the resource are pseudo-orthogonal between different virtual cells. VCI reduces RS interference between different transmitting points to ensure the reliable demodulation of reference signal. The CoMP cooperating sets (macro cell, micro cell) can be configured according to the location of the UE. With UE feedback, Rel.11 introduced the concept of the channel state information of the process. UE periodic and non-periodic feedback are both based on the channel state information process. Interference measurement and channel quality indicator are newly defined to support the accurate measurement of the channel state information.

Rel.12 introduced inter-cell CoMP, namely the distributed CoMP with non-ideal backhaul. New signaling interaction between X2 are introduced, such as RSRP measurement reports, etc. Cooperative or mute transmission scheme is supported. The non-ideal backhaul scenes in R12 can be applied to the macro-micro scenario.

In this paper, the principle of different strategies of CoMP are presented, the challenge for backhaul transmission and delay is analyzed. The UL JR and DL CS algorithm and flow are discussed in Sect. 3. Followed by simulations and field trial in heterogeneous network. Finally, a conclusion is given in the end.

2   Principle and Challenge of COMP

CoMP involves several possible coordinating schemes among the access points. Firstly, CoMP can be applied both in downlink and uplink. Secondly, there are inter-site and intra-site CoMP according to the cooperating objects. Multiple sectors of one base station (eNB in 3GPP LTE terminology) can cooperate in intra-site COMP, whereas inter-site COMP involves multiple eNBs. Furthermore, downlink CoMP can be classified as Joint Processing, including Joint Transmitting (JT), Dynamic Point Selection or Blanking (DPS/DPB), and also Coordinated Scheduling or Beamforming (CS/CB). With JT, multiple cells transmit identical data by using the same Resource Block (RB), which improves the performance of reception, working as diversity gain from MIMO. With DPS, multiple points share the same data like JT, but the data is sent by one cell with best channel quality while other cells are muted. CS CoMP allocates different RB to cell-edge UEs to avoid interference, and CB CoMP utilized beamforming technology to transmit orthogonal resources. Similarly, Uplink CoMP has Joint Reception (JR) in the uplink scheduling and coordination beamforming. The following Fig. 1 details the principles of these CoMP techniques [8].

Unlike ICIC or eICIC, CoMP uses not only the frequency and time domain resources, but also the spatial domain, known as a fast interference coordination. Therefore, the fast-changing UE channel information must be reported during each scheduling took place. UEs measure their Channel State Information (CSI) and report to eNB, which includes Channel Quality Indicator (CQI), Rank indicator (RI) and Precoding Matrix Indicator (PMI). For this purpose, eNB gives UEs instruction on which cell's CSI are be measured by using particular RB, CSI-Reference Signal.

The delay requirements for transmitting CSI and data are strict, especially for JT and JR, the CQI and users' data must be shared between the transmission and

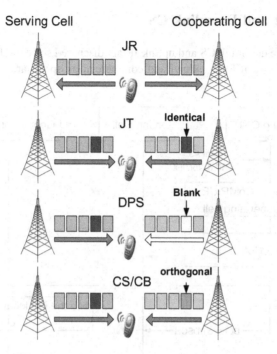

Fig. 1. The framework of PCI Self-Configuration

Table 1. Information and delay Requirement for CoMP

Type	Information requires	Magnitude of delay	Resource domain
JR	User data, JR scheduling, Reference signal configuration information, CSI	100 μs, Ideal backhaul	Frequency, Spatial
JT	User data, JT scheduling, Precoding information, CSI	100 μs, Ideal backhaul	Frequency, Spatial
DPS	Scheduling, Reference signal configuration information, Precoding information, CSI	Ms, Ideal backhaul	Frequency, Time, Spatial
CS/CB	Scheduling, CSI	Ms, Ideal backhaul	Frequency/Spatial

coordinating cells in TTI level, which brings the transmission network a challenge not only on delay but also in bandwidth. The requirement of different strategy are shown in Table 1. In this situation, the use of fiber link for Common Public Radio Interface (CPRI) is necessary in commercial cases.

3 Uplink JR and Downlink CS

In this section, the downlink CS and uplink JR are discussed since the JT and DPS have demanding requirement for backhaul and commercial prospects are uncertain.

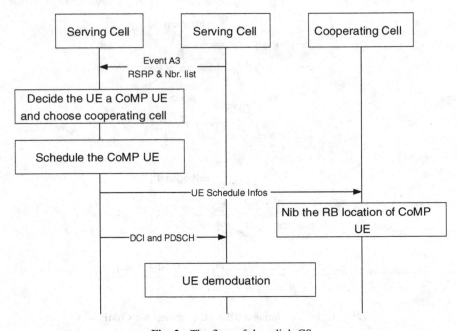

Fig. 2. The flow of downlink CS

Centralized and distributed algorithm both works for downlink CS according to the location where algorithm executes. For centralized CS, scheduling and channel state information of each cell are transmitted to a centralized control network element, from where the user downlink resource are uniformly managed. However, the scheduling decision are made by the serving cell in a distributed algorithm, the serving cell manages the downlink CS resource based on the scheduling and CSI information transmitted from cooperative cells. As shown in Fig. 2:

(1) UE measures RSRP according to A3 message, and then report the neighbor cell list and neighbor relations to the serving cell.
(2) UE serving cell determined whether trigger CoMP or not, and identify the coordinated neighbor cells.
(3) Serving cell schedule UE and sent the scheduling information to the coordinated cells.
(4) Coordination cells avoid scheduling the identical CoMP-used RB resource while scheduling their own UE.
(5) Serving cell transmits the scheduling information and data to the CoMP UE.
(6) UE perform demodulation.

One key parameter to trigger the algorithm is SINR of UE, particularly the UE at the cell edge receive strong interference. The situation can be formulized as:

$$SINR = \frac{P_s}{I_{MAX} + I_{others} + N}, \tag{1}$$

Where P_s is signal power of serving cell, I_{MAX} is the strongest signal power from neighbor cell, I_{others} is the interference power from other cells combine, and N is noise power. After CS executing, the $SINR$ is improved:

$$SINR' = \frac{P_s}{I_{others} + N}, \tag{2}$$

For cell edge UE, usually the rate is limited by the interference, and the I_{MAX} always much stronger than the other interference and noise, $I_{MAX} \gg I_{others} + N$. Therefore, after applying CS, $SINR'$ is improved significantly, $SINR' \gg SINR$.

On the other hand, uplink JR chooses two (or more) qualified joint cells to cooperate. For uplink UE, twice the antennas participate in joint receiving. The performance gain obtain from two aspects.

- Joint gain: the signal sent by the UE at the cell edge (overlapping the two cells) can be simultaneously received by different cell antennas. Enhance joint reception received higher signal quality.
- Interference restrain gain: UE in cooperative cell is selected for joint process who receives the interference from the UE at cell edge. The UL CoMP joint progress restrain the interference to obtain the interference restrain gain.

In a cellular network, the JR gains distributes in different regions. As shown in Fig. 3, different colors indicates different types of gain:

- Light blue: Joint gain obtain from intra-BBU
- Yellow and orange: Joint gain obtain from inter-BBU
- Dark blue: Interference restrain gain from intra-BBU

Data combining takes place after receiving from separate antennas. The position of combining affects the process complexity, inter-cell transmission, and the performance gain. Extra physical processes are needed to support uplink JR. The procedure is showed in Fig. 4.

Additional physical layer operation under CoMP:

(1) Channel estimation is not only required by source UE, but also by neighbor UE.
(2) Soft information of CoMP UE is obtain after two user equalization.
(3) CoMP UE performance gain by combining the soft information.

In general, transmitting the original time domain or frequency domain I/Q signal obtains higher performance gain at the expense of higher complexity and transmit bandwidth. On the other hand, transmitting the soft bit data after demodulation require lower resource but acquire lower gain.

4   Field Trial and Measurement Results

In this section, field trials are performed to further investigate the performance of typical strategies of CoMP. Some large-scale field trial has been carried out in urban area in Shang Hai. Table 2 shows the basic field trial parameters setting. There are 19 cells on 7 sites involved in CoMP JR. The average distant of every two sites are 350 M, and the antenna height are 20 m. The system bandwidth is 20 MHz and the UE is carried on a measurement vehicle with speed of 10 km/h and the data traffic transmitted from the beginning till the end [8].

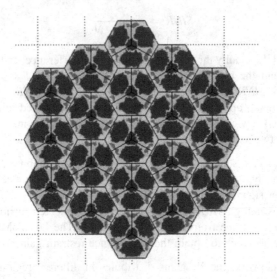

Fig. 3. The gain area of uplink JR (Color figure online)

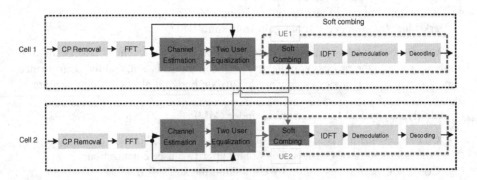

Fig. 4. The physical procedure of uplink JR

When the signal quality is poor, the uplink transmission between different shared TPs is launched. TPs measured the UE's channel quality conditions; According to the measurement results, the network selects strong signal node joint data reception, in this

Table 2. Parameters setting

PARAMETERS	ASSUMPTION
Cellular layout	19 cells of 7 eNBs
User layout	Circling, 1 users
Cell radius	350 m
BS Transmit Power	46 dBm
Carrier frequency	2 GHz
Band Width	20 MHz

case, the threshold is set to 6 dB. As shown in Fig. 5, non-cooperative area is compared to the cooperative area. Because of cooperation, the average through put gains about 20%–30%. In certain areas, even higher gains over 50% were observed.

Under HetNet scene, the main difference between HetNet CoMP and homogeneous Network is the power difference [9, 10]. The strategy can be the same in HetNet except the RSRP or other parameters should meet the threshold to activate CoMP.

In addition, the downlink CS feature is tested in a heterogeneous network scene. In this scenario, the UE moves from the center to the edge of the interfered cell (micro cell), which is under the signal coverage of the macro cell completely. The results in Fig. 6 shows that RSRP and the through puts are reducing gradually since the UE moves outward. The throughput decreases and finally approaches zero (shown as blue part); However, when the DL CS is on, the UE throughput raised in different degree (shown as red part), because of the coordination between the macro and micro cells. The user information is transmitted to the macro cell, macro cell schedule different RBs to avoid the interference. When the user moves to the cell edge, the avoidance of interference has become more evident, the cell spectral efficiency is further improved, and gain significantly compared to the downlink CS off.

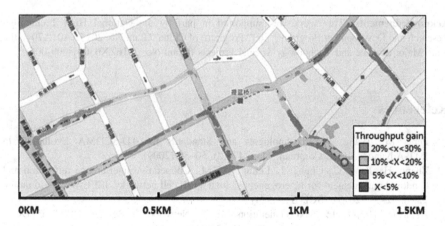

Fig. 5. The uplink JR gains in Shang Hai

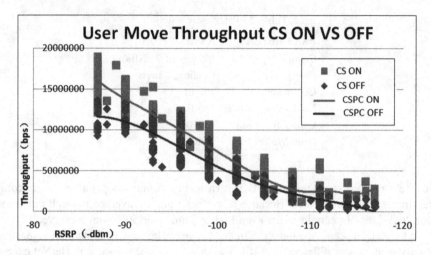

Fig. 6. The throughput of downlink CS on vs. off (Color figure online)

5 Conclusions

In this article, the overview of CoMP technology is given, the difference of five types of CoMP, both in uplink and downlink, are analyzed. Challenges and requirement for backhaul and delay are presented. As JT and DPS are regarded as the uneconomic methods for network operator, the UL JR and DL CS are more likely be applied in the future network. Therefore, flow and strategies of UL JR and DL CS are discussed in detail. Field trial has been taken both in homogeneous and heterogeneous network to support the conclusion. The trial results shows that under a certain condition, CoMP can provide higher bitrates and improve cell edge user rate significantly, and system spectral efficiency is improved as well.

Acknowledgements. This work was supported in part by the National High Technology Research and Development Program ("863" Program) of China (Grant No. 2015AA01A704), the State Major Science and Technology Special Projects (Grant No. 2016ZX03001-009-006).

References

1. Peng, M., Wang, W.: Technologies and Standards for TD-SCDMA Evolutions to IMT-Advanced. IEEE Commun. Mag. **47**(12), 50–58 (2009)
2. Zhang, H., Jiang, C., Cheng, J., Leung, V.C.M.: Cooperative interference mitigation and handover management for heterogeneous cloud small cell networks. IEEE Wirel. Commun. **22**(3), 92–99 (2015). (SCI, IF: 6.524)
3. Akhavan, H., et al.: Next Generation Mobile Networks – Beyond HSPA EVDO - Whitepaper Technical report, NGMN Ltd. (2006). www.ngmn.org

4. Zhang, H., Jiang, C., Beaulieu, N.C., Chu, X., Wen, X., Tao, M.: Resource allocation in spectrum -sharing OFDMA femtocells with heterogeneous services. IEEE Trans. Commun. **62**(7), 2366–2377 (2014). (SCI, IF: 1.979)
5. Zhang, H., Jiang, C., Beaulieu, N., Chu, X., Wang, X., Quek, T.: Resource allocation for cognitive small cell networks: a cooperative bargaining game theoretic approach. IEEE Trans. Wirel. Commun. **14**(6), 3481–3493 (2015). (SCI, IF: 2.762)
6. Aziz, D., Sigle, R.: Improvement of LTE handover performance through interference coordination. In: IEEE 69th Vehicular Technology Conference, 2009. VTC Spring 2009 (2009)
7. Zhu, H., Wang, J.: Chunk-based resource allocation in OFDMA systems - Part I: chunk allocation. IEEE Trans. Commun. **57**(9), 2734–2744 (2009)
8. European Cooperative in the Field of Science and Technical Research EURO-COST231, Urban transmission loss models for mobile radio in the 900 and 1800 MHz bands, rev. 2, The Hague, September 1991
9. Zhang, H., Liu, H., Jiang, C., Chu, X., Nallanathan, A., Wen, X.: A practical semi-dynamic clustering scheme using affinity propagation in cooperative picocells. IEEE Trans. Veh. Technol. **64**(9), 4372–4377 (2015). (SCI, IF: 2.642)
10. Zhang, H., Jiang, C., Mao, X., Chen, H.-H.: Interference -limit resource optimization in cognitive femtocells with fairness and imperfect spectrum sensing. IEEE Trans. Veh. Technol. **65**(3), 1761–1771 (2016). (SCI, IF: 2.642)

Design and Analysis of Economic Games

Revenue Sharing of ISP and CP in a Competitive Environment

Nari Im, Jeonghoon Mo$^{(\boxtimes)}$, and Jungju Park

Yonsei University, Seoul, Korea
{nariim,j.mo,jungju.park}@yonsei.ac.kr

Abstract. We considered a revenue sharing problem between a content provider (CP) and a Internet service provider (ISP) when two ISPs competes with each other. ISPs can provide a piracy monitoring service, which may increase the profit of CP, to incentivize CP to collaborate with one of them. We modeled the problem as a multi-stage game and characterized an equilbrium content price, piracy monitoring level, and revenue sharing ratio. We found a condition in which ISP and CP may collaborate even under competition. We also provide numerical results.

Keywords: Revenue sharing · Content piracy · ISP competition

1 Introduction

The importance of Internet for contents delivery is getting more and more important as more people consume them with Internet. The number of subscribers of Netflix, the largest online streaming service provider, has reached 69 Million (Q3, 2015) according to statista.com [1]. The number of youtube user is more than 1 Billion and 4 Billion video views are consumed every day [2]. According to Cisco, global IP traffic has increased more than five-fold in the past five years, and the wireless traffic growth rate is expected to be 61% per year from 2013 to 2018 [3].

Such a high traffic growth put a burden on Internet infrastructure, network upgrades in both backbone and access are needed. Internet service providers all over the world have invested in 3G and LTE wireless access network for last several years. For example, three Korean service providers (SKT, KT, and LGU+) spent between $6B and $8B per year during 2011–2013 [4]. To reflect such a burden on investment, the major US ISPs including AT&T, Comcast, TWC, and Verizon claimed that extra regulation would threaten new investment and innovation on network upgrades [5].

The debate between ISPs and CPs have been on-going under the name of *network neutrality*. One of the main issues is about how to share the investment costs for network upgrade in a reasonable manner. CP side argues that it is necessary for the new innovation and fair competition. The other side argues that network neutrality can hinder the proper development of network infrastructure and deployment of high-quality services.

© ICST Institute for Computer Sciences, Social Informatics and Telecommunications Engineering 2017
J. Cheng et al. (Eds.): GameNets 2016, LNICST 174, pp. 113–121, 2017.
DOI: 10.1007/978-3-319-47509-7_11

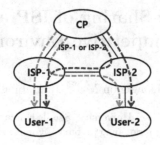

Fig. 1. Internet ecosystem

Researchers have considered possibility of content charge by ISPs for recovering investment cost [6–9]. Kamiyama (2014a, 2014b) considered a content charge system that ISPs charge a fee for each content delivery using a 3 stage Stackelberg model. In [8,9], authors explored a possibility of revenue sharing with a means of piracy monitoring. The ISPs provide a piracy monitoring service to CPs to increase the demand for legitimate contents. According to [10], the traffic of illegal contents represents about 23.8 % of total Internet traffic. For example, the popular Netflix show "House of Cards" season three was downloaded illegally approximately 682,000 times within the first 24h of being available [11]. If such service of ISP can help increasing the profit of CPs, CPs can be more willing to collaborate with ISPs in revenue sharing.

In this work, we extend the results of [8,9] to a competitive situation in which there are two ISPs. When one ISP asks for profit sharing, CPs can switch to another ISP who does not require the sharing. Therefore, introduction of profit sharing may not always be a good solution to the ISP. In our model, we assume that only one of the ISPs provides the piracy monitoring service, while the other does not, to understand the impacts of competition on the profit sharing behavior.

The rest of the paper is organized as follows: In Sect. 2, we explain the details of our game models including players, payoff functions, and sequence of games. Section 3 derives the best responses of users, CP, and ISP-1, and finds each player's strategy. In Sect. 4, we show the simulation results of model. Finally, we conclude the paper in Sect. 5.

2 System Model

We consider an Internet ecosystem that consists of a CP, two ISPs, and a set of N users as shown in Fig. 1. CP provides its contents to the users using networks of the ISPs. It contracts with one of ISPs, say ISP-i, for the network access and pays a_i per unit traffic for delivery of contents for $i = 1, 2$. We assume that two ISPs are in a peering relationship and do not need to settle for the exchange.

The two ISPs are different in that ISP 1 requests revenue sharing to CP while ISP 2 does not. If CP contracts with ISP 1, it shares γ fraction of its revenue with the ISP on top of the access fee. In return for the revenue sharing, ISP 1 provides a strengthened piracy monitoring service to protect copyrighted contents to be spread over the Internet. For example, a technique such as DPI (deep packet inspection) can be used to implement such a protection service. As the ISP deliberately involves in contents protection, it is more difficult for users to infringe copyrighted content over ISP-1's network.

Users pay a content price p to CP for getting the contents. The price could be different depending on which ISP is chosen by CP. On the other hand, they can get the contents in an illegal way without paying the price[1].

User Utility: A user of type v has the following utility function:

$$u_v = \begin{cases} v - p, & \text{if a type } v \text{ user buys a legal content;} \\ (1 - \alpha)v - \beta, & \text{if a type } v \text{ user acquires an illegal content;} \\ 0, & \text{otherwise,} \end{cases} \quad (1)$$

where v is the willingness to pay for the contents, p is the content price, α is quality degradation factor between 0 and 1, and β is the cost for copyright infringement. If a type v user purchases a contents, its net-utility is $v - p$. If he acquires it in an illegal way, the value of illegal copy is smaller than a legitimate one by αv. The second term β models efforts or costs for acquiring an illegal copy. When ISP 1 provides strengthen piracy protection, the value of β increases.

CP Profit: The profit of CP depends on contents price p, profit sharing ratio γ, access fee a per unit content, and protection level β. Let ϕ_i be the profit function of CP when it contracts with ISP i, for $i = 1, 2$. Then, it is given by:

$$\phi_i(p; \gamma, \beta, a) = D(p, \beta)((1 - \gamma)p - a), i = 1, 2, \quad (2)$$

where $D(p, \beta)$ is a demand function for legitimate contents.

Profit of the ISPs: Let π_i be the profit function of ISP i, then it is given by:

$$\pi_i(\beta; p, \gamma, a) = (a + \gamma p)D(p, \beta) - c(\beta), i = 1, 2, \quad (3)$$

where β is the piracy protection level, p is the content price, γ is the revenue sharing ratio, and a is the access fee. Here, $aD(p\beta)$ is the revenue from the access charge, $\gamma pD(p, \beta)$ is the revenue from the profit sharing and $c(\beta)$ is the cost for maintaining piracy protection level β. We assume that $c(\beta)$ is nondecreasing function of β. As ISP 2 does not request profit sharing, γ of ISP-2 is 0, and we assume that β of ISP-2 is constant to be $\underline{\beta}$. Hence,

$$\pi_2(\underline{\beta}; p, 0, a) = aD(p, \underline{\beta}) - c(\underline{\beta}).$$

We further assume that the access fee a is the same for two ISPs as the access network market is competitive.

---

[1] For example, P2P service such as bitTorrent provides a way to getting a contents without proper payment.

Sequence of Game: We consider a game between ISP-1 and CP. Even though ISP-2 is in the model, ISP-2 does not have any strategies to control unlike ISP-1. ISP-1's controls profit sharing ratio γ, piracy protection level β. CP's needs to determine which ISP to choose and the price p for the content. We model the game as multi-stage sequential game as follows:

1. ISP-1 and CP negotiate profit sharing ratio γ_1.
2. Given profit sharing ratio γ_1, ISP-1 determines the monitoring level β_1.
3. Given (γ_i, β_i) of ISP-i, $i = 1, 2$, CP selects one of the ISPs to maximize its profit and determines its content price p.
4. Let $\beta = \beta_1$ if ISP-1 is chosen; or $\beta = \beta_2$, otherwise. Given (p, β), users determines its behavior among three possibilities: buying level content, acquiring illegal content, or doing nothing.

3   Analysis of Best Responses and Equilibrium

3.1   User Behaviors

To maximize their utilities of (1), users select one of three options: (a) buying a legitimate contents, (b) downloading an illegal content, and (c) doing nothing. A user of type v makes a legal purchase if $v - p \geq (1 - \alpha)v - \beta$ and $v - p \geq 0$ or if $v \geq v_0 := \max(\frac{p-\beta}{\alpha}, p)$. Similarly, he downloads an illegal content if $(1 - \alpha)v - \beta \geq v - p$ and $(1 - \alpha)v - \beta \geq 0$ or $\frac{\beta}{1-\alpha} \leq v \leq \frac{p-\beta}{\alpha}$.

If the distribution function of user type is $F(\cdot)$, then the demand $D(p, \beta)$ for legal contents can be expressed as

$$D(p, \beta) = 1 - F(v_0) = 1 - F\left(\max(\frac{p-\beta}{\alpha}, p)\right). \tag{4}$$

Here, we normalized the maximum demand to be 1 without loss of generality.

If $\frac{p-\beta}{\alpha} \leq p$, only legal purchase can happen; otherwise, legal or illegal contents downloads coexist. To see this, note that the first condition implies $\beta \geq (1-\alpha)p$. The cost β of piracy is so high it is better off for users to buy legal contents. On the other hand, if $\frac{p-\beta}{\alpha} \geq p$ or $\max(\frac{p-\beta}{\alpha}, p) = \frac{p-\beta}{\alpha}$, both legal and illegal contents coexist.

We can limit our attention to $\beta \leq p(1 - \alpha)$ or $\max(\frac{p-\beta}{\alpha}, p) = \frac{p-\beta}{\alpha}$ because an equilibrium always exists in the low β regime. When there is no piracy users, increasing β no longer helps the ISP but costs more. Hence, the ISP does not increase β more than $p(1-\alpha)$. Hence, the above demand function can be rewritten into:

$$D(p, \beta) = 1 - F\left(\frac{p - \beta}{\alpha}\right).$$

If the consumer type v is uniformly distributed on the continuum of $[0, \bar{v}]$, where $\bar{v}$ is the maximum willingness to pay, the demand becomes a linear function:

$$D(p, \beta) = 1 - \left(\frac{p - \beta}{\bar{v}\alpha}\right). \tag{5}$$

3.2 . Strategy of CP

CP determines the optimal content price p^* and chooses one of the ISPs to contract with. Let p_i^* be the optimal price on the condition that CP contracts with ISP-i. Then, we have the following proposition on optimal p_i^*.

Proposition 1. *Assume that CP contracts with ISP-i. Given monitoring level β_i and profit sharing rate γ_i, the optimal content price p_i^* of CP is:*

$$p_i^* = \begin{cases} p_i^{1*} := \frac{\alpha\bar{v}+\beta_i}{2} + \frac{a}{2(1-\gamma_i)}, & \text{if } \beta \leq \beta_c; \\ p_i^{M*} := \frac{\beta_i}{1-\alpha}, & \text{oherwise.} \end{cases} \quad (6)$$

where $\beta_c = \frac{1-\alpha}{1+\alpha}(\alpha\bar{v} + \frac{a}{1-\gamma_1})$.

Sketch of proof: As the profit function (2) of CP is concave in p_i, applying the first order condition gives the desired results. ∎

With the optimal content price p_i^*, CP determines which ISP network to use for a larger profit. Note that for any monitoring level β_i and profit sharing rate γ_i, $\phi_i(p_i^*; \gamma_i, \beta_i, a) \geq 0$. Then, we compare the profit from contracting each ISP. In case of contracting with ISP-1, the profit function of CP is given as:

$$\phi(p_1^*; \gamma_1, \beta_1, a) = \begin{cases} \frac{-(1-\gamma_i)}{\bar{v}(1-\alpha)^2}(\beta_i - \bar{v}(1-\alpha))(\beta_i - \frac{a(1-\alpha)}{1-\gamma_i}), & \text{if } \beta \leq \beta_c; \\ \frac{(1-\gamma_i)^2}{4\alpha\bar{v}(1-\gamma_i)}(\beta_i - (\frac{a}{1-\gamma_i} - \alpha\bar{v}))^2, & \text{otherwise,} \end{cases} \quad (7)$$

where $\beta_c = \frac{1-\alpha}{1+\alpha}(\alpha\bar{v} + \frac{a}{1-\gamma_1})$. We can also derive the profit function of CP contracting with ISP-2 by plugging $\gamma_2 = 0$ and $\beta_2 = \beta_L{}^2$ as:

$$\phi(p_2^*; \gamma_2, \beta_2, a) = \frac{(\alpha\bar{v} - a)^2}{4\alpha\bar{v}}, \quad (8)$$

where $p_2^* = \frac{\alpha\bar{v}+a}{2}$.

Shape of $\phi(p_1^*; \gamma_1, \beta_1, a)$: First, we characterize the shape of $\phi(p_1^*; \gamma_1, \beta_1, a)$. There are three different increasing/decreasing patterns of $\phi(p_1^*)^3$ as a function of β in terms of γ_1 as shown in Fig. 2. The three plots correspond to the three cases, (A) $\gamma \leq 1 - \frac{a}{\alpha\bar{v}}$; (B) $1 - \frac{a}{\alpha\bar{v}} < \gamma \leq 1 - \frac{a}{\bar{v}}$; and (C) $\gamma > 1 - \frac{a}{\bar{v}}$, respectively. Due to space limitaition, we omit the detailed derivations.

ISP Selection: CP selects ISP-1 if $\phi(p_1^*) \geq \phi(p_2^*)$. Otherwise, it chooses ISP-2. After thorough analysis, we have following proposition and observations. It turns out that the access fee plays an important role in ISP selection. We skip the analysis due to space limitation.

Proposition 2. *If $a < \alpha^2\bar{v}$ and $1 - \alpha \leq \gamma_1 \leq 1 - \frac{a^2}{\alpha^2\bar{v}^2}$, contracting with ISP-2 always provides a higher profit to CP than doing with ISP-1.*

---

2 For the simplicity of analysis, we assume that $\beta_L = 0$.
3 We will use $\phi(p_1^*)$ and $\phi(p_1^*; \gamma_1, \beta_1, a)$, interchangeably, for the sake of readability.

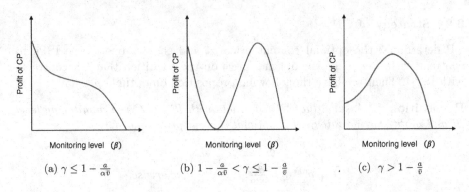

Fig. 2. Three different shapes of $\phi(p_1^*)$

Proposition 2 says that selecting ISP-2 is always better off than selecting ISP-1 with a small access fee a and a large profit sharing rate γ_1, regardless of monitoring level β_1. Otherwise, we can always find some monitoring level with which contracting with ISP-1 is more beneficial.

Let $\mathcal{B} := \{\beta_1 | \phi(p_1^*) \geq \phi(p_2^*), \phi(p_1^*) \geq 0\}$.

Observation

1. CP tends to be better off with ISP-1 as an access fee a increases.
2. When γ_1 is small, there exists $\tilde{\beta}$ such that $\mathcal{B} = \{0 \leq \beta_1 \leq \tilde{\beta}\}$.
3. When γ_1 is large, there exist $\check{\beta}$ and $\hat{\beta}$ such that $\mathcal{B} = \{\beta | 0 < \check{\beta} \leq \beta_1 \leq \hat{\beta}\}$

Observation 1 describes the influences of the access fee changes. With a small access fee, contracting with ISP-1 is less likely to be more beneficial to CP. In other words, given a profit sharing level γ_1, the higher access fee, the larger domain of β_1 that provides a higher profit. In addition, we observe some counterintuitive findings.

In general, it is easy to think that a monitoring level and a profit sharing rate increases or decreases together. However, the above two observations shows a different views. To the former, though the profit sharing rate is small, the monitoring level is bounded below by $\check{\beta}$ in order for ISP-1 to be chosen. Similarly, to the latter, the monitoring level is supposed to be bounded above by $\hat{\beta}$. These mean that there can be a minimum and a maximum levels of a monitoring level for ISP-1 to be chosen by CP. For more details, refer the technical report.

3.3 Strategy of ISP

In this section, we find the optimal monitoring level β_1^* of ISP-1 that maximizes its profit $\pi_1(\beta_1; p_1^*, \gamma_1)$. We assumed that $c(\beta) := \kappa\beta^2$ where κ is a positive constant. Plugging (5) and (6) into the profit function (3) of ISP-1, we have

$$\pi_1(\beta_1; p_1^*, \gamma_1) = \begin{cases} -\frac{\gamma_1 + \kappa\bar{v}(1-\alpha)^2}{\bar{v}(1-\alpha)^2}\beta_1^2 + \frac{(\bar{v}\gamma_1 - \alpha)}{\bar{v}(1-\alpha)}\beta_1 + a, & \text{if } \beta_1 \leq \beta_c; \\ \frac{\gamma_1 - 4\kappa\alpha\bar{v}}{4\alpha\bar{v}}\beta_1^2 + \frac{\alpha\bar{v}\gamma_1 + a}{2\alpha\bar{v}}\beta_1 + \frac{(a(2-\gamma_1) + \alpha\bar{v}\gamma_1(1-\gamma_1))(\alpha\bar{v}(1-\gamma_1) - a)}{4\alpha\bar{v}(1-\gamma_1)^2}, & \text{otherwise.} \end{cases}$$

Table 1. Increase/decrease of ISP profit function and the potential optimal monitoring levels without competition*

Cases		Range of β_1		
		0	β_c	∞
[1] $a \geq 4\alpha\bar{v}^2 k$				
[1-1]	$\gamma_1 < 4\alpha\bar{v}k,\quad \beta^{1*} < 0$	(0) ↘	↘	↘ ↘
[1-2]	$\gamma_1 < 4\alpha\bar{v}k,\ 0 \leq \beta^{1*} < \beta_c$	↗ (β^{1*}) ↘	↘	↘ ↘
[1-3]	$\gamma_1 < 4\alpha\bar{v}k,\ \beta_c \leq \beta^{1*}$	↗	↗ (β_c) ↘	↘
[1-4]	$4\alpha\bar{v}k \leq \gamma_1 < \frac{a}{\bar{v}},$	↗	↗ (β_c) ↘	↘
[1-5]	$\frac{a}{\bar{v}} \leq \gamma_1,\qquad\qquad \beta^{M*} < \beta_c$	↗	↗ (β_c) ↘	↘
[1-6]	$\frac{a}{\bar{v}} \leq \gamma_1,\qquad\qquad \beta_c \leq \beta^{M*}$	↗	↗	↗ (β^{M*}) ↘
[2] $a < 4\alpha\bar{v}^2 k$				
[2-1]	$\gamma_1 < a/\bar{v},\qquad \beta^{1*} < 0$	(0) ↘	↘	↘ ↘
[2-2]	$\gamma_1 < a/\bar{v},\ \ 0 \leq \beta^{1*} < \beta_c$	↗ (β^{1*}) ↘	↘	↘ ↘
[2-3]	$\gamma_1 < a/\bar{v},\ \ \beta_c \leq \beta^{1*}$	↗	↗ (β_c) ↘	↘
[2-4]	$a/\bar{v} \leq \gamma_1 < 4\alpha\bar{v}k,\ \beta_c \leq \beta^{1*},\qquad \beta_c \leq \beta^{M*}$	↗	↗	↗ (β^{M*}) ↘
[2-5]	$a/\bar{v} \leq \gamma_1 < 4\alpha\bar{v}k,\qquad \beta^{1*} < 0,\ \beta_c \leq \beta^{M*}$	(0) ↘	↘	↗ (β^{M*}) ↘
[2-6]	$a/\bar{v} \leq \gamma_1 < 4\alpha\bar{v}k,\ 0 \leq \beta^{1*} < \beta_c, \beta_c \leq \beta^{M*}$	↗ (β^{1*}) ↘	↘	↗ (β^{M*}) ↘
[2-7]	$a/\bar{v} \leq \gamma_1 < 4\alpha\bar{v}k,\ \beta^{1*} \geq \beta_c,\qquad \beta^{M*} < \beta_c$	↗	↗ (β_c) ↘	↘
[2-8]	$a/\bar{v} \leq \gamma_1 < 4\alpha\bar{v}k,\ \beta^{1*} < \beta_c,\qquad \beta^{M*} < \beta_c$	↗ (β^{1*}) ↘	↘	↘ ↘
[2-9]	$4\alpha\bar{v}k \leq \gamma_1,\qquad\qquad \beta^{M*} < \beta_c$	↗	↗ (β_c) ↘	↘
[2-10]	$4\alpha\bar{v}k \leq \gamma_1,\qquad\qquad \beta_c \leq \beta^{M*}$	↗	↗	↗ (β^{M*}) ↘

*Each figure within a bracket is a critical point, except (0) which is the lower bound.

The problem of ISP-1 can be formulated as the following constrained optimization problem:

$$\max_{\beta_i} \pi_1(\beta_1; p_1^*, \gamma_1) \tag{9}$$

$$\text{sub. to} \quad \beta_1 \in \mathcal{B}, \tag{10}$$

where $\mathcal{B} = \{\beta_1 | \phi(p_1^*) \geq \phi(p_2^*), \phi(p_1^*) \geq 0\}$. The constraint is needed due to competition with ISP-2. If ISP-1 determines β_1 such that $\beta_1 \notin \mathcal{B}$, as CP selects ISP-2, its profit becomes zero.

It turns out that profit function $\pi_1(\beta_1; p_1^*, \gamma_1)$ is either decreasing or unimodal in most cases except [2–4] and [2–5] of Table 1[4]. We found the optimal β_1^* for 16 cases of Table 1. However, the solution does not incorporate the competition constraint. As it becomes too complicated to find the analytic result of the optimization problem (9)–(10) of ISP-1, we rely on numerical studies for the ISP-1's optimal strategy.

3.4 Negotiation of Profit Sharing Rate

The negotiation of revenue sharing ratio γ is a difficult issue in reality and is beyond the scope of this paper. What we would like to see is whether there exists $\gamma > 0$ such that both ISP and CP can be happier than when $\gamma = 0$.

---

[4] We omit details due to lack of space.

One possible theoretical approach is to use the concept of Nash bargaining solution [12], which can be expressed as

$$\gamma^* = \arg\max_{\gamma} \phi(p_1^*) \times \pi(\beta_1^*; p_1^*, \gamma_1)$$

4   Numerical Study

In this section, we made a numerical study to present the possibility of a ISP-CP collaboration. We assumed that the access fee a is 0.3; the maximum willingness to pay $\bar{v}$ is 10; the quality degradation α is 0.1; and the constant of the monitoring cost κ is 0.2. It is supposed to reflect a market situation with an access fee of middle level $(\alpha^2\bar{v} \leq a < \alpha\bar{v})$ and a low quality degradation level in which we can avoid extreme cases for the levels of access fee and reflect the ease of piracy in real world.

We analyzed the cases; the Internet access fee a is .1, .3, and .5 where $\alpha^2\bar{v} = .1$ and $\alpha\bar{v} = 1$. Figure 3 shows the changes of ISP and CP profits in terms of the profit sharing rates γ from 0 to .5. For all cases, the trends of their profit functions were the same that the profit of CP is unimodal and the profit of ISP increases. The existence a positive profit sharing rate until which the profit functions of ISP and CP commonly increase implies shows a great possibility that the profit sharing of CP can be beneficial to not only ISP, but itself.

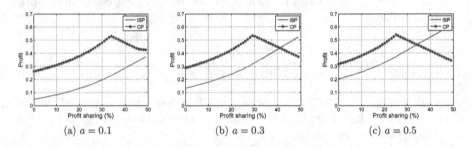

(a) $a = 0.1$ (b) $a = 0.3$ (c) $a = 0.5$

Fig. 3. Profit of ISP-1 and CP for different values of α

5   Conclusions

In this paper, we studied a possible collaboration between ISP and CP when two ISPs compete with each other. We formulated the problem as a multi-stage game model of which players are two ISPs and one CP. While ISP-1 provides a piracy monitoring service, ISP-2 does not. In return, ISP-1 requests revenue sharing to CP. CP determines its content price and selects one of the them to maximize its profit.

We characterized equilibrium strategies of ISP and CP. We also found a condition in which ISP and CP may collaborate with each other. When an access fee is small and a revenue sharing ratio is high enough, CP has no incentive to participate in the revenue sharing. Otherwise, (either an access fee is high or a revenue sharing ratio is low enough) there exists a monitoring level that ISP and CP can collaborate.

Further characterization of piracy monitoring level remains as future work as well as that of revenue sharing ratio. In addition, extension of the game model to multiple CPs will be pursued.

Acknowledgment. This work was supported in part by the National Research Foundation of Korea Grant funded by the Korean Government (NRF-2012R1A1A2007890).

References

1. http://www.statista.com/statistics/250934/quarterly-number-of-netflix-streaming-subscribers-worldwide/
2. Smith, C.: By the numbers: 120+Amazing Youtube statistics. http://expanded-ramblings.com/index.php/youtube-statistics/
3. Cisco Visual Index Report. http://www.cisco.com/c/en/us/solutions/service-provider/visual-networking-index-vni/index.html
4. Money Today Article, Febuary 2013 (in Korean). http://news.mt.co.kr/mtview.php?no=2013020516041781122
5. Reardon, M.: Broadband CEOs to FCC: Observation of strains: We're not a utility. In: CNET(2011), 13 May 2004
6. Kamiyama, N.: Effect of content charge by ISPs in competitive environment. In: 2014 IEEE Network Operations and Management Symposium (NOMS), pp. 1–9. IEEE, May 2014a
7. Kamiyama, N.: Feasibility analysis of content charge by ISPs. In: 2014 26th International Teletraffic Congress (ITC), pp. 1–9. IEEE, September 2014b
8. Park, J., Mo, J.: ISP and CP revenue sharing and content piracy. ACM SIGMETRICS Perform. Eval. Rev. **41**(4), 24–27 (2014)
9. Park, J., Im, N., Mo, J.: ISP and CP collaboration with content piracy. In: 2014 IEEE International Conference on Communication Systems (ICCS), pp. 172–176. IEEE, November 2014
10. Filloux, F.: Piracy is part of the digital ecosystem (2012). http://www.co.uk/technology/2012/jan/23/monday-note-piracy-sopa
11. Trefis Team. Is Piracy A Serious Threat To Netflix? Forbes, 17 March 2015
12. Nash, J.: The bargaining problem. Econometrica JSTOR **18**(2), 155–162 (1950)

An Incentive Mechanism for P2P Network Using Accumulated-Payoff Based Snowdrift Game Model

Ruoxi Sun[1,2], Wei Li[3], Haijun Zhang[4,5(✉)], and Yong Ren[1]

[1] Department of Electronic Engineering, Tsinghua University, Beijing 100084, China
srx13@mails.tsinghua.edu.cn, reny@tsinghua.edu.cn
[2] China Transport Telecommunications and Information Center,
Beijing 100011, China
[3] The State Radio Monitoring Center, Beijing 100037, China
liwei@srrc.org.cn
[4] The University of British Columbia, Vancouver, B.C. V6T 1Z4, Canada
dr.haijun.zhang@ieee.org
[5] National Mobile Communications Research Laboratory,
Southeast University, Nanjing, China

Abstract. Cooperation between participators has played a very important role in P2P Network. Whereas, in contradiction to the original design philosophy of P2P file sharing system, it is difficult to guarantee the cooperation of these participators and hard to maintain a high stability of the network due to the selfishness of people without behavior constraints. In this paper, we propose a novel incentive mechanism using Accumulated-Payoff Based Snowdrift Game (APBSG) model to improve frequency of cooperation for P2P network. The performance analysis of this model and simulation results show that APBSG can reduce the sensitivity of cooperation to the selfishness of nodes, which promotes the cooperative behavior in P2P network to a large extent. Meanwhile, we reveal the relationship between the degree distribution and the frequency of cooperation by analyzing APBSG features under small-world and scale-free network. The result suggests that we can adopt different strategies according to degrees of nodes to achieve better stability for P2P network.

Keywords: P2P · Snowdrift game · Incentive mechanism · Scale-free network

1 Introduction

P2P file sharing system has occupied an increasingly important position in Internet applications [1]. However, unlike the traditional HTTP service, each participator is not only a downloader, but also an uploader in P2P network, which requires more cooperation between participators. Nevertheless, each participator is selfish since everyone wants to download more files from others with few contributions. The file sharing system will eventually tend to crash if a high frequency of cooperation can not be well maintained in P2P network. In fact, these selfish "free riders" [2,3] will result in at least two problems:

© ICST Institute for Computer Sciences, Social Informatics and Telecommunications Engineering 2017
J. Cheng et al. (Eds.): GameNets 2016, LNICST 174, pp. 122–132, 2017.
DOI: 10.1007/978-3-319-47509-7_12

- The participators refuse to share their files, and the network resource will be seriously under-utilized.
- The participators download with an unlimited rate, which will lead to the waste of bandwidth and the instability of system.

Hughes et al. [3] found in 2005 that 85 % of Gnutella users are free riders. There are many studies [4] suggest that traditional P2P file sharing systems suffer from free riding due to lacking of an effective incentive mechanism.

Therefore, in order to solve these two problems, an effective mechanism, which can guarantee both the downloading efficiency of participators and the proportion of cooperation behavior should be designed [4]. If each participant not only complies with the rules and restraints downloading rate, but also is willing to share files with others, the P2P system will reach an evolutionary balance. On the contrary, those free riders who do not abide by the rules may get more payoffs in the short term, but then they will be revenged by the system and their payoff would be sharply reduced.

The cooperation mechanism is very complex in such a dynamic system. Since cooperation is ubiquitous all around the real world ranging from biological systems to economic and social systems. Game theory is considered to be an important approach and a powerful framework to solve these problems. In this paper, to emphasize the core issue of cooperation, we consider this mechanism as a snowdrift game [5]. Our goal is to design a simple but effective incentive mechanism to guarantee the cooperation of the participators in the entire network which runs in an efficient and continuous way. An incentive mechanism using Accumulated-Payoff Based Snowdrift Game (APBSG) model is presented, which makes the whole network achieve a high frequency of cooperation and thereby increase the network stability in different conditions. We also analyze the relationship between the frequency of cooperation and degree of nodes, which offers another way to improve the stability of the network.

Similar with other complex networks, P2P network has small-world and scale-free properties. Thus, we conduct our simulation and analysis separately on small-world and scale-free networks instead of on random networks or lattices. Simulation results show that APBSG can reduce the sensitivity of cooperation to the selfishness of nodes, which greatly promotes the cooperative behavior in P2P network.

The rest of the paper is organized as follows. In Sect. 2, we briefly explain the snowdrift game and topological characteristics of P2P network. A cooperation incentive mechanism for P2P network is given in Sect. 3 and simulation results will be presented in Sect. 4. Finally, Sect. 5 gives conclusions and future work.

2   Related Work

2.1   Snowdrift Game

Snowdrift Game describes the situation that involves two drivers both want to go home, but they are trapped on opposite sides of a snowdrift. Each of them can make the choice of staying in the car (defect-D) or shoveling the snow (cooperation-C) to clear a path. If they both shovel the snow from the opposite sides which means they both choose cooperation strategy, both of them can obtain the payoff R by driving home and sharing the labor cost of shoveling snow. If one of them choose to stay in the car and the other shovels snow, the cooperator gets a payoff S and the defector yields the highest payoff T. However if both stay in the car, they cannot go home and obtain the minimum payoff P. Thus, we have $T > R > S > P$, and the payoff of the defect and cooperation behaviors can be formulated in a payoff matrix as following:

$$\begin{array}{cc} & C\ D \\ \begin{array}{c} C \\ D \end{array} & \begin{pmatrix} R\ S \\ T\ P \end{pmatrix} \end{array} \tag{1}$$

[6,7] shows that snowdrift game is an evolutionary game and both sides will eventually converge to the evolutionary strategy. Meanwhile, game theory has been widely utilized in modeling various networks in [8,16].

2.2 Topological Characteristic of P2P Network

Many recent works have shown that a lot of complex networks have the same or similar features in the real world. Actually, the P2P network is one type of complex networks and has the same features [17–20]. The most typical features are that they all have a small average path length, a large clustering coefficient and a long tail on degree distribution. This makes the P2P network differ a lot from the lattice or random graph. Two of the most famous models for simulation are the Small-World network proposed by Watts and Strogatz [19] in 1998 and Scale-free network which is proposed in 1999 by Barabási and Albert [20].

Figure 1 shows a small-world network and a scale-free network, which both have 50 nodes. Figure 1(a) shows a small-world network with an average degree of 10 which derives from the nearest-neighbor-coupled network. In this network, every edge is cut the connection in a probability of p with its neighbors and reconnected up to another node. When $p = 0$, the network is a regular nearest-neighbor-coupled network, and the increase of p makes the existence of "short-cut" in the network, which brings a sharp reduce of average path length. But the clustering coefficient remains very large due to the fact that most edges are still connected to the neighbor nodes. When $p = 1$, the network becomes a random network.

A different strategy is taken for scale-free network with an average degree of 13 which is shown in Fig. 1(b). It derives from a full-connected network with m_0 nodes, then adds one node at each step, from which m edges ($m \leq m_0$) are added to the existent nodes in a probability of $p_i = \dfrac{k_i}{\sum\limits_{j=1}^{N} k_j}$. [20] shows that

degrees of nodes in network generated in such a mechanism meets the power-law distribution.

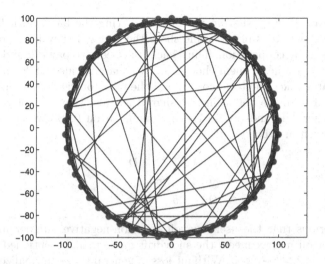

Fig. 1. Small-world network and Scale-free network.

3   Cooperation Incentive Mechanism

3.1   Assumption

Assume that the topology of realistic P2P network is in accordance with the small-world and scale-free networks mentioned above. In this paper, we conduct our simulation based on this assumption. Before modeling, we need to make some assumptions relating to P2P network:

- In this model, each node represents a participant in P2P network, and the edge between two nodes means there is a relationship of uploading and downloading between them. To simplify the model, we consider the topology of network is static, which means that all the nodes will not withdraw from the network and no new node will be added in, and also the connection between nodes will not be changed.
- Each node needs to choose the cooperation or defect strategy to game with its neighbor nodes for every round and gets a certain payoff which is the summary of payoffs when gaming with every neighbor nodes.
- Each node will choose the strategy to get a maximum payoff for next round.

This game strategy is configurable, i.e., the system can be pre-configured to control the downloading and uploading for each node. In fact, specific client software is necessary when downloading, and the strategy will be setup inside the software.

3.2   Accumulated-Payoff Based Snowdrift Game

Consider a P2P network of N nodes, for each pair of nodes with connection, normal utilization by downloading brings users a payoff b, meanwhile they must pay an extra cost c by sharing files and observing the bandwidth limitations. Whereas, the two sides in the game share the cost c if they both comply with this agreement by cooperation. Thus, each user needs to pay $c/2$ and get a total payoff $b - c/2$. If one of them chooses not to share its files while the other do, the cooperator takes the total cost c, and the defector yields the highest payoff b. Whereas if both are not willing to share their files they cannot get anything from the other side and obtain the minimum payoff 0. So we can obtain the following payoff matrix:

$$
\begin{array}{cc}
 & \begin{array}{cc} C & D \end{array} \\
\begin{array}{c} C \\ D \end{array} & \begin{pmatrix} b - c/2 & b - c \\ b & 0 \end{pmatrix}
\end{array}
\tag{2}
$$

It is obvious that the user's payoff is a non-negative number in any case. Thus the payoff matrix meets the snowdrift game model mentioned previously since $b > b - c/2 > b - c > 0$. Without loss of generality, we normalize the payoff matrix by defining $b - c/2 = 1$ and $r = c/2$, where r for the payoff ratio. The normalized payoff matrix is as follows:

$$
\begin{array}{cc}
 & \begin{array}{cc} C & D \end{array} \\
\begin{array}{c} C \\ D \end{array} & \begin{pmatrix} 1 & 1 - r \\ 1 + r & 0 \end{pmatrix}
\end{array}
\tag{3}
$$

The effects of each node's accumulated payoff have not received enough attention in the study of snowdrift game on complex networks in previous works. Since individuals always make decisions based on the payoffs they got in the past time, we decide to construct a model based on accumulated payoff.

Here defines two variables $\pi_C(i)$ and $\pi_D(i)$ to indicate the accumulated payoff from the initial state to current round of node i for cooperation and defect strategy respectively. Strategy in kth round is determined by $\pi_C(i)$ and $\pi_D(i)$. Also, $P_C(i)$ and $P_D(i)$ is defined as follows:

$$
P_C(i) = \frac{\pi_C(i)}{\pi_C(i) + \pi_D(i)}
\tag{4}
$$

$$
P_D(i) = \frac{\pi_D(i)}{\pi_C(i) + \pi_D(i)}
\tag{5}
$$

Obviously, $P_C(i) + P_D(i) = 1$, so we define $P_C(i)$ as the probability of choosing the strategy of cooperation in the kth round, also $P_D(i)$ for defect. Without loss of generality, we initialize both $\pi_C(i)$ and $\pi_D(i)$ to 1 before the start of the game.

In our model, the strategy of choosing cooperation or defect for the next round is not determined by the strategy or payoff in a specific round, but by

---

Algorithm 1. Accumulated-Payoff Based Snowdrift Game

1: **Input** r, N
2: **Initialize** the initial strategies of all nodes $s[1...N]$
3: **Initialize** $\pi_C(i) = \pi_D(i) = 1$
4: **for** $k = 1$ to the number of round **do**
5:    **for** node $i = 1$ to N **do**
6:       **Update** $\pi_C(i)$ and $\pi_D(i)$
7:       **Calculate** $P_C(i)$ and $P_D(i)$ using Eqs. (4) and (5)
8:       **Generate** a random number R ranging from 0 to 1
9:       **if** $R > P_C(i)$ **then**
10:          $s(i) =$ Defect
11:       **else**
12:          $s(i) =$ Cooperation
13:       **end if**
14:    **end for**
15: **end for**

---

the proportion of the accumulated payoff of all previous game rounds. Hence, we call this model as Accumulated-Payoff Based Snowdrift Game (APBSG).

The algorithm is summarized by the following pseudo code:

For example, before the kth round begins, if a node i has $\pi_C(i) = 60$ and $\pi_D(i) = 40$, it will choose cooperation strategy in a probability of 0.6 and defect strategy of 0.4. To achieve this effect, we generate a random number R ranging from 0 to 1. The node will choose defect strategy if R is greater than 0.6, otherwise choose cooperation. Another benefit for this algorithm is that it is easy to promote the generation of cooperation behavior, i.e., cooperation is probably to emerge in a network where all the participators take defect as their initial strategy because of the principle that strategy for each round is not absolute, but in a certain probability. With the game advancing, more and more nodes will tend to an evolutionary equilibrium to choose the strategy in a certain probability to maximize their payoffs.

4 Simulation

As mentioned above, our goal is to guarantee the downloading efficiency and the stability of the P2P network. This mainly relies on the frequency of cooperation, which is the proportion of nodes taking cooperation strategy while the system gets an evolutionary balance in the network. We define f_c as the frequency of cooperation. It can be easily seen that f_c ranges from 0 to 1. When $f_c = 0$, there's no nodes choosing cooperation strategy, while $f_c = 1$ means that all the nodes have chosen cooperation strategy in the network.

In our simulation, the performance metric is f_c, and we observe how f_c changes as a function of different parameters both in the small-world network and scale-free network.

4.1   APBSG on Small World Network

Firstly, we investigate APBSG on small-world network. Simulation is carried out for a population of $N = 2500$ nodes. Figure 2 shows the results where f_c as a function of parameter r.

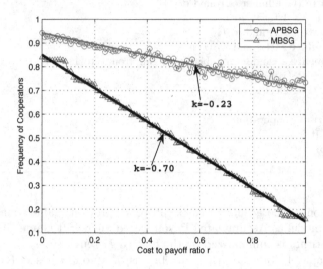

Fig. 2. f_c as a function of r in small-world network with average degree of 200. The network size is 2500. f_c for each simulation is obtained by averaging from time step $t = 1000$ to 5000 where the system has reached a steady state.

As a comparison, the result of Memory-Based Snowdrift Game (MBSG) [21] is also shown. It can be seen from Fig. 2 that f_c linearly decreases while the payoff ratio r increases both in APBSG and MBSG, i.e., the greater the r is, the more payoff a user will obtain from others while taking defect strategy, which inhibits the cooperation in the network. In particular, when $r = 1$, for a node i, no matter which strategy it takes, it gains nothing as long as it's neighbors take defect strategy. In this case, the frequency of cooperation drops to below 0.2 under MBSG. It is a very bad situation where more than 80 % of the participators do not offer uploading service or over-occupied downloading bandwidth in a P2P network, the network has in fact degenerated to the traditional HTTP service. Those nodes taking cooperation strategy can be considered as the servers and the other users download files from these servers.

Although f_c also declines in APBSG, the decrease speed is very slow. f_c still remains at up to 0.7 when $r = 1$.

The increase of r leads to a high payoff when choosing the defect strategy, thus the value of r represents the degree of a rational selfish individual. A rational individual's selfishness is to maximize its payoff as far as possible, and a unilateral defect always brings a greater payoff. Therefore, we defined $\theta(r)$ as the sensitivity of the frequency of cooperation $f_C(r)$ to nodes' selfishness when payoff ratio is r:

$$\theta(r) = \left| \lim_{\Delta x \to 0} \frac{f_C(r + \Delta r) - f_C(r)}{\Delta r} \right| \tag{6}$$

The simulation result is an approximately straight line for both APBSG and MBSG in Fig. 2, thus $\theta(r)$ equals the absolute value of the slope of fitting curve when r ranges from 0 to 1. We can calculate that slope of fitting line in APBSG is -0.23, and -0.7 for MBSG, so we can see that $\theta_{APBSG}(r) = 0.23$ and $\theta_{MBSG}(r) = 0.7$ in these simulation conditions. Thus it can be concluded that the sensitivity for frequency of cooperation to individuals' selfishness in APBSG is only about 1/3 of that in MBSG and P2P network using APBSG is better in stability than in MBSG when the condition changes. Due to the fact that a strong punishment will be taken when the one node chooses defect behavior, although one can get a temporarily high payoff in one round, it will immediately change its strategy from defect to cooperation to gain a maximum payoff.

Figure 3 shows how frequency f_c of cooperation changes with different degrees in small-world network which has an average degree of 200 when $r = 0.2$. When the degree ranges from 180 to 200, i.e., those nodes having middle degrees are typical "strategy swingers" who will choose a cooperation strategy in one round and maybe defect for next round. Although they have a high average percentage of cooperation, they almost never choose one strategy continuously to the end. In contrast, those nodes that have smaller or larger degrees tend to be "pure cooperators". They will always adhere to the cooperative behavior no matter what strategy their neighbors take.

According to this, if the node distribution of P2P network is similar to small-world network, we can appropriately reduce the degree of nodes that have intermediate degree by controlling its connection to others using client software to promote a more emergence of cooperative strategies, which will go a long way towards improving the stability of P2P network.

4.2   APBSG on Scale-Free Network

Going beyond small-world, we also investigate the APBSG on scale-free network. Figure 4 shows the simulation results on the Barabási-Albert network.

Result in Fig. 4 shows that similar to the results in the small-world network, the frequency of cooperation linearly decreases when payoff ratio r increases in scale-free network. However, there're still about 80 % nodes choose cooperation strategy while in small-world it's about 70 % when $r = 1$, and the sensitivity $\theta(r)$ of the frequency of cooperation is 0.15, which is lower than that in small-world network. This indicates that scale-free network is more likely to promote emergence of cooperation than small-world network because of the wider range of degree distribution in scale-free network.

Figure 5 shows how the frequency of cooperation changes with the different degrees in scale-free network when $r = 0.2$. It can be easily seen that only the nodes with larger degrees tend to be "pure cooperators", while most of the other nodes tend to be "strategy swingers", which is very different from that in small-world. Thus we can increase the heterogeneity of nodes for P2P network to improve the frequency of cooperation and guarantee the stability of the network.

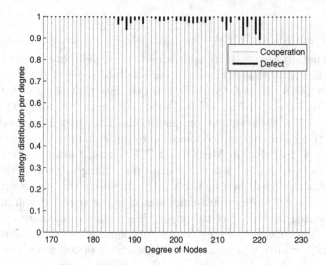

Fig. 3. Distributions of strategies in small-world network. Cooperators and defectors are denoted by gray bars and black bars respectively. Each bar adds up to a total fraction of 1 per degree, the gray and black fractions being directly proportional to relative percentage of respective strategy for each degree. Those nodes which have middle degrees will choose the cooperation strategy in one round and maybe defect for next round. In contrast, those having smaller or larger degrees tend to adhere to the cooperative behavior no matter what strategy their neighbors take.

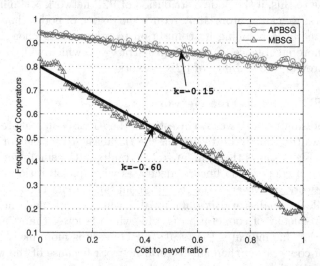

Fig. 4. f_C as a function of r in scale-free network whose size is 2500. f_C for each simulation is obtained by averaging from time step $t = 1000$ to 5000 where the system has reached a steady state.

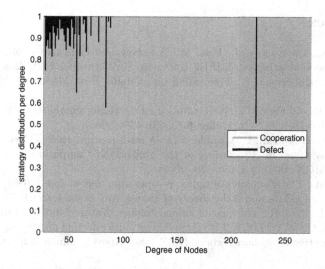

Fig. 5. Distributions of strategies in scale-free network. Cooperators and defectors are denoted by gray bars and black bars respectively. Each bar adds up to a total fraction of 1 per degree, the gray and black fractions being directly proportional to relative percentage of respective strategy for each degree. Those nodes which have smaller degrees will choose the cooperation strategy in one round and maybe defect for next round. In contrast, those having larger degrees tend to adhere to the cooperative behavior no matter what strategy their neighbors take.

5   Conclusion

In this paper, we have investigated the cooperative behavior in the P2P file sharing system and found that it's essentially a snowdrift game. A novel incentive mechanism called Accumulated-Payoff Based Snowdrift Game is adapted to guarantee a high proportion of cooperation and maintain a continuing stability of the P2P network. The results show that APBSG can reduce the sensitivity of cooperation to the selfishness of nodes, which greatly promotes the cooperative behavior in P2P network. Meanwhile, the results also give a relationship of frequency of cooperation and the degrees of nodes. Nodes with larger or smaller degrees promote the emergence of cooperative behavior in small-world network, while in scale-free networks, the larger degree nodes tend to be pure cooperators.

In current work, the topology of network is static, and this assumption is true for some cases. In future work, we will investigate the incentive mechanism in the situation when nodes can withdraw from or add dynamically in to the P2P network.

Acknowledgment. This work was supported by the NSFC China under projects 61371079, 61471025, 61271267 and 91338203, and the Open Research Fund of National Mobile Communications Research Laboratory, Southeast University (No. 2016D07).

References

1. Lua, E.K., Croowcroft, J., Pias, M.: A survey and comparition of peer-to-peer overlay network schemes. J. IEEE Commun. Surv. Tutorial **7**(2), 72–93 (2005)
2. Adar, E., Huberman, B.: Free riding on gnutella. First Monday **5**(10), 305–314 (2000)
3. Hughes, D., Goulson, G., Walkerdine, J.: Free riding gnurella revisited: the bell tolls? IEEE Distrib. Syst. Online **6**(6), 276–277 (2005)
4. Saroiu, S., Gummadi, P., Gribble, S.D.: A measurement study of peer-to-peer file sharing systems. In: Proceeding of the Multimedia Computing and Networking 2002 (MMCN 2002), pp. 156–170 (2002)
5. Sugden, R.: The economics of rights, co-operation and welfare (1986)
6. Smith, M.: Evolution and the theory of games. Am. Sci. **64**(1), 41–45 (1976)
7. Smith, M., P. G. R.: The logic of animal conflict. Nature 246(5427), 15–18 (1973)
8. Jiang, C., Chen, Y., Liu, K.J.R.: Data-driven route selection and throughput analysis in cognitive vehicular networks. IEEE J. Sel. Areas Commun. **32**(11), 2149–2162 (2014)
9. Jiang, C.: Graphical evolutionary game for information diffusion over social networks. IEEE J. Sel. Topics Signal Process. **8**(4), 524–536 (2014)
10. Jiang, C.: Evolutionary dynamics of information diffusion over social networks. IEEE Trans. Signal Process. **62**(17), 4573–4586 (2014)
11. Jiang, C., Chen, Y., Gao, Y., Liu, K.J.R.: Joint spectrum sensing and access evolutionary game in cognitive radio networks. IEEE Trans. Wireless Commun. **12**(5), 2470–2483 (2013)
12. Jiang, C., Chen, Y., Liu, K.J.R.: Distributed adaptive networks: a graphical evolutionary game theoretic view. IEEE Trans. Signal Process. **61**(22), 5675–5688 (2013)
13. Jiang, C., Chen, Y., Yang, Y., Wang, C., Liu, K.J.R.: Dynamic chinese restaurant game: theory and application to cognitive radio networks. IEEE Trans. Wireless Commun. **13**(4), 1960–1973 (2014)
14. Zhang, N.B.X.C.X.W.H., Jiang, C., Quek, T.Q.: Resource allocation for cognitive small cell networks: A cooperative bargaining game theoretic approach. IEEE Trans. Wireless Commun. **14**(6), 3481–3493 (2015)
15. Zhang, H.: Resource allocation with interference mitigation in ofdma femtocells for co-channel deployment. EURASIP J. Wireless Commun. Networking 89, (2012)
16. Jiang, C., Chen, Y., Liu, K.J.R.: Multi-channel sensing and access game: Bayesian social learning with negative network externality. IEEE Trans. Wireless Commun. **13**(4), 2176–2188 (2014)
17. Albert, R., Barabási, A.-L.: Statistical mechanics of complex networks. Rev. Mod. Phys. **74**, 47–97 (2002)
18. Boudec, J.Y.L., Vojnovic, M.: Perfect simulation and stationarity of a class of mobility models. In: Proceedings of IEEE INFOCOM 2005, pp. 72–79 (2005)
19. Watts, D.J., Strogatz, S.H.: Collective dynamics of 'small-world' networks. Nature **393**, 440–442 (1998)
20. Barabási, A.L., Albert, R.: Diameter of the world-wide web. Nature **401**, 130–131 (1999)
21. Wang, W.X., Ren, J., Chen, G., Wang, B.H.: Memory-based snowdrift game on networks. Phys. Rev. E **74**, 56–113 (2006)

Joint Power Control and Subchannel Allocation for D2D Communications Underlaying Cellular Networks: A Coalitional Game Perspective

Yanjie Dong$^{(\boxtimes)}$, Md. Jahangir Hossain, and Julian Cheng

School of Engineering, The University of British Columbia,
Kelowna, BC V1V 1V7, Canada
yanjie.dong@alumni.ubc.ca, {julian.cheng,jahangir.hossain}@ubc.ca

Abstract. A coalition based joint subchannel and power allocation approach is studied to improve the performance of device-to-device (D2D) communication underlaying cellular networks with uplink spectrum sharing. To exploit the spectrum reuse gain, we formulate the problem as a coalition formation game. Furthermore, a distributed coalition formation algorithm is devised to assist D2D pairs in joining or leaving a coalition. During the coalition formation process, we introduce an iterative power control method. By using this method, D2D pairs can evaluate their current coalition with D2D sum rate maximization and cellular user equipment protection. Numerical results are provided to corroborate the proposed studies.

Keywords: Coalitional game theory · Power control · Device-to-device communication

1 Introduction

Nowadays, the demand for wireless internet access witnesses a huge increment. Cisco Systems, Inc. estimates that the wireless data traffic will continue to grow exponentially and reach over 24 exabytes per month in 2019. Device-to-device (D2D), a type of proximity communication, has been proposed to work underlaying existing cellular network for spectrum efficiency improvement. In a D2D pair, under the control of evolved NodeB (eNB), user equipments (UEs) communicate with each other through direct link instead of resorting to eNB's assistance. Under this network architecture, the spectrum band can be utilized simultaneously by both D2D pairs and traditional cellular pairs, and the D2D links can exploit the spectrum reuse gain without any hardware investment. As a result, D2D communication is involved as a key component in LTE-Advanced systems (Doppler et al. 2009, Lei et al. 2012) and in the fifth generation communication systems (Boccardi et al. 2014, Tehrani et al. 2014).

© ICST Institute for Computer Sciences, Social Informatics and Telecommunications Engineering 2017
J. Cheng et al. (Eds.): GameNets 2016, LNICST 174, pp. 133–146, 2017.
DOI: 10.1007/978-3-319-47509-7_13

Despite the benefits, D2D communication may also cause two types of inter-ference: interference to primary cellular user equipments (CUEs) and interfer-ence to other D2D pairs that use the same frequency band. Current literature mainly focus on the problems of power control (Yu et al. 2009a,b 2009, Xing and Hakola 2010, Dong et al. 2016), subchannel allocation (Yu et al. 2011, Xu et al. 2013, Xu et al. 2014) and interference management (Janis et al. 2009, Xu et al. 2010) by considering both types of interference. For example, the authors in (Yu et al. 2009) showed that proper power control can coordinate the interference to maximize the sum rate. In (Xu et al. 2013), a reverse iterative combinatorial auction based method was proposed to efficiently allocate subchannel resource to the D2D pairs, which operates in the downlink period of CUEs. These meth-ods manage the interference from D2D to CUEs when D2D pairs operate in the downlink period.

On the other hand, fewer works have considered the interference management that occurs in the uplink transmission. Indeed, the uplink interference manage-ment is a more challenging issue because the interference control process is left to multiple UEs instead of the single eNB. Moreover, existing works (Yu et al. 2009a,b Xing and Hakola 2010, Yu et al. 2011, Janis et al. 2009, Xu et al. 2010) only considered the resource allocation and interference management under a restricted scenario where only one D2D pair coexists with one CUE.

Some recent literature (Min et al. 2011, Wang et al. 2013, Feng et al. 2013, Li et al. 2014) considered a more practical scenario with multiple D2D pairs or multiple CUEs. The authors in (Min et al. 2011) studied a case in which multiple CUEs coexist with a D2D pair and proposed a location based interference man-agement approach. The proposed approach defined an interference limited area for D2D pair where CUEs cannot share the spectrum with the D2D pair. The authors in (Wang et al. 2013) assumed that a D2D pair can reuse the channels of multiple CUEs. They developed a suboptimal algorithm to jointly allocate the transmission power of CUEs and the D2D pair such that the throughput of the D2D pair is maximized, and the QoS of CUEs are guaranteed. The authors in (Feng et al. 2013) formulated the resource allocation problem as a system throughput maximization problem with the assumption that the resource of a CUE can be shared at most by one D2D pair. The authors in (Li et al. 2014) introduced coalitional game theory to model the subchannel allocation in D2D communication underlaying uplink cellular network. However, they did not study the topic of power control for D2D pairs which may help D2D pairs further exploit the spectrum reuse gain.

In this paper, we formulate the problem of joint subchannel and power allo-cation for D2D enabling system as a coalition formation game. For the devised game model, a distributed coalition formation algorithm is proposed, where each D2D pair can make decision to leave or join a coalition. Within a specific coali-tion, each D2D pair tries to optimize the its utility via power control. Here, the utility of each D2D pair is formulated as difference between achieved spectrum efficiency with the priced power cost. Each D2D pair evaluates its satisfaction level on the current coalition based on its achieved utility. Our contributions are summarized as follows.

- We propose a coalition formation algorithm for D2D pairs to select the subchannel. We prove the coalition formation algorithm converges to a Nash stable partition.
- For a specific coalition, we derive a distributed iterative power control algorithm to mitigate the interference on CUEs and interference among D2D pairs. We also discuss the convergence issue for the power control algorithm.

Simulation results illustrate that the proposed scheme can increase the sum rate of both CUE and D2D pairs. Meanwhile, the proposed algorithm can also reduce the unnecessary coalition switch operations.

The rest of this paper is organized as follows. In Sect. 2, the system model is described. In Sect. 3, we model the D2D pair coalition formation game to allocate the subchannels, and a distributed algorithm is proposed. Moreover, we discuss the power control in each coalition. Simulation results are given in Sect. 4. Section 5 concludes our works.

2   System Model

We consider the uplink of orthogonal frequency division multiple access (OFDMA) based wireless network, where an eNB is located at the center of the cell and multiple UEs are distributed uniformly within the cell. This network contains two types of UEs, i.e., M CUEs and N D2D pairs where $N > M$. Let $\mathcal{M} = \{1, 2, \ldots, M\}$ and $\mathcal{N} = \{1, 2, \ldots, N\}$ denote the CUE set and the D2D pairs set, respectively. Moreover, the distance between two UEs in a D2D pair satisfies the constraint of D2D communication. We assume all CUEs utilize orthogonal subchannels and D2D pairs share the subchannels with CUEs. The subchannel assignment for CUEs is fixed, and multiple D2D pairs can share one subchannel with the CUE simultaneously to improve the system spectrum efficiency.

Figure 1 illustrates the existing interference under the above network setting in uplink period. We can see that there are two types of interference, e.g., interference among D2D pairs and interference between CUE and D2D pairs. For example, let us consider a case where the 1st and the 2nd D2D pairs share the same subchannel with CUE c_1. Thus, the corresponding D2D receivers d_1^r and d_2^r are exposed to the interference from CUE c_1. While the eNB receives interference from d_1^t and d_2^t which are the transmitters of the 1st and the 2nd D2D pairs respectively. Meanwhile, there exists interference between the 1st and the 2nd D2D pairs. CUE c_2 and the 3rd D2D pair use orthogonal subchannels, and as a result, they do not interfere with each other.

As the number of D2D pairs increases, both types of interference will become more severe. Therefore, if interference is not managed properly, the potential gain in spectral efficiency obtained by spectrum sharing will be wiped out. Motivated by this fact, we focus on power control and subchannel assignment for D2D pairs.

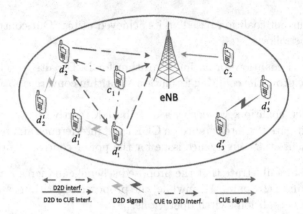

Fig. 1. Multiple D2D pairs coexist with multiple cellular users.

We denote $X = \left[x_i^k\right]_{N \times M}$ as the subchannel assignment matrix where x_i^k takes either 1 or 0 to indicate whether the subchannel of kth CUE is assigned to ith D2D pair or not, $i \in \mathcal{N}$ and $k \in \mathcal{M}$. We allow a D2D pair to use only one subchannel, that is, $\sum_{k \in \mathcal{M}} x_i^k \leq 1$. Based on these assumptions, the received signal at the eNB of CUE $k \in \mathcal{M}$ and the signals at receiver of ith D2D pair underlying CUE k can be, respectively, written as

$$y_k = \sqrt{p_k H_k} s_k + \sum_{i \in D_k} x_i^k \sqrt{p_i G_i^k} s_i + n_k \tag{1}$$

and

$$z_i^k = \sqrt{p_i h_{i,i}^k} s_i + \sqrt{p_k g_i^k} s_k + \sum_{j \in D_k \backslash \{i\}} x_j^k \sqrt{p_j h_{j,i}^k} s_j + n_i^k \tag{2}$$

where s_i and p_i are the signal and the transmit power of the ith transmitter, $i \in \mathcal{M} \cup \mathcal{N}$; the terms H_k and G_i^k denote the channel gain of CUE k and the interference gain between D2D pair i to CUE k, respectively; $h_{i,j}^k$ is the channel gain between the transmitter of D2D pair i to the receiver of D2D pair j underlying CUE k; the set D_k represents the D2D pairs share the subchannel of CUE k, $D_k \subset \mathcal{N}$, and D_k can be empty; n_k and n_i^k are the additive white Gaussian noise of CUE k and D2D pair i underlying CUE k with power N_0.

3   Interference Mitigation as a Coalition Formation Game

In this section, we first present the coalition formation game formulation. Then, we analyze the power control issue in a specific coalition. At last, we propose a distributed coalition formation algorithm.

3.1   Coalitional Game in Partition Form

In the studied network, there are M CUEs and N D2D pairs, where D2D pairs choose to share the subchannels with CUEs to enhance the network sum rate throughput. We allow multiple D2D pairs operating on the same subchannel of a CUE to form cooperative group, i.e., *coalition*. As a result, we denote the *coalition partition* as $\pi = \{D_1, D_2, \ldots D_M\}$, where $\bigcup_{k=1}^{M} D_k = \mathcal{N}$, $D_k \cap D_m = \phi$, $\forall k, m = 1, 2, \ldots, M$ and $k \neq m$. Note that $D_k = \phi$ means no D2D pair reuses the subchannel of CUE k.

Based on above analysis, we can denote the received SINR for CUE k and D2D pair i as

$$\Gamma_k = \frac{p_k H_k}{\sum_{i \in D_k} x_i^k p_i G_i^k + N_0} \tag{3}$$

and

$$\gamma_i^k = \frac{p_i h_{i,i}^k}{p_k g_i^k + \sum_{j \in D_k \setminus \{i\}} x_j^k p_j h_{j,i}^k + N_0}. \tag{4}$$

Furthermore, we can calculate the throughput of UEs by the Shannon formula $r = \log_2 (1 + \text{SINR})$.

Note that with the increase of D2D pairs in the coalition D_k, the interference among the CUE and D2D pairs will increase. Thus, D2D pairs will deviate from their current coalition to join another coalition for their throughput improvement. This motivates us to employ the coalitional game theory (Saad et al. 2009) to formulate the coalition switch mathematically. In this paper, we formulate the joint power and subchannel allocation as a coalition formation game in partition form with nontransferable utility.

Definition 1. *A coalition formation game with non-transferable utility (NTU) for joint power and subchannel allocation in D2D communication network is defined by a pair $(\mathcal{N}, V)$ where $\mathcal{N}$ is the set of players[1] and V is a mapping such that for every coalition $D_k \subset \mathcal{N}$, $k = 1, 2, \ldots, M$, $V(D_k)$ is a closed convex subset of $\mathbb{R}^{D_k}$ that contains the payoff vectors that players in D_n can achieve.*

Denoting by $v_i (D_n)$ the payoff of D2D pair i in coalition $D_n \in \pi$, thus the coalition value set is defined as

$$V(D_k) = \left\{ \boldsymbol{v}(D_k) \in \mathbb{R}^{D_k} \,\middle|\, v_i (D_k), i \in D_k \right\} \tag{5}$$

and the payoff of each D2D pair is

$$v_i (D_k) = r_i \left(p_i^*, p_{-i}^* \right), \forall i \in D_k \tag{6}$$

where p_i^* is the transmit power of D2D pair i, p_{-i}^* is the transmit power of other D2D pairs belonging to the same coalition as i, i.e. $-i \in D_n \setminus \{i\}$. Both of them will be determined by the power control scheme afterwards. The NTU property indicates the payoff for each D2D pair depends on the joint actions of all the D2D pairs in the coalition (Saad et al. 2009).

---

[1] We use the same set of D2D pairs as all the D2D pairs join the formulated game.

3.2 Power Control Within a Specific Coalition

After forming a coalition, all D2D pairs in this coalition, say $D_k \in \pi$, work cooperatively to maximize their sum rate. Meanwhile, all the D2D pairs are punished as they cause excessive interference on the CUE. That means, D2D pairs can evaluate a coalition with both sum rate maximization and CUE protection from interference. In the proposed game model, the punishment is linear to the total transmission power of D2D pairs on subchannel k. Thus, we can derive the optimal power $\boldsymbol{p}^* (D_k) = \left(p_1^*, p_2^*, \ldots, p_{|D_k|}^* \right)^2$ of D2D pairs by solving the following optimization problem

$$\max_{\boldsymbol{p}(D_k)} \sum_{i \in D_k} \log_2 \left(1 + \gamma_i^k \left(\boldsymbol{p} \left(D_k \right) \right) \right) - \lambda \sum_{i \in D_k} p_i \tag{7a}$$

$$s.t. \ 0 \le p_i \le p_{\max}, \forall i \in D_k \tag{7b}$$

where λ is a fixed linear price factor; constraint (7b) gives the power range.

From (3)–(4), we notice that the optimization problem (7a)–(7b) is nonconvex, which can be complex to solve. As a result, we consider a low-complexity distributed iterative method to find a local optimum point. Then, optimization programming (7a)–(7b) is replaced by

$$\max_{p_i \in \boldsymbol{p}(D_k)} \ \log_2 \left(1 + \gamma_i^k \left(\boldsymbol{p} \left(D_k \right) \right) \right) - \lambda p_i$$
$$s.t. \ 0 \le p_i \le p_{\max}, \forall i \in D_k. \tag{8}$$

Generally, the optimization problem (8) is convex and can be solved using a standard method (Boyd and Vandenberghe 2004). As a result, we develop Algorithm 1 to generate a sequence of transmit power for D2D pairs in D_k.

Algorithm 1. Iterative Power Control Algorithm (IPC)

1: All D2D initialize their power $p_i(t) = 0, \forall i \in D_k$, iteration count t, power price λ and maximum iteration number MAX.
2: **repeat**
3:     $t := t + 1$
4:     Transmitter of D2D pair i estimates the interference-plus-noise level, i.e., the denominator of (4), $\forall i \in D_k$
5:     Transmitter of D2D pair i estimates the channel gain $h_{i,i}^k$ using the received signal power of control packet, $\forall i \in D_k$
6:     Transmitter of D2D pair i get the transmit power of tth iteration p_i^t by solving (8), $\forall i \in D_k$
7: **until** $t > MAX$ or $\left\| p_i^t - p_i^{t-1} \right\| \le \epsilon, \forall i \in D_k$

Following proposition provides a sufficient condition of the convergence properties of (7a)–(7b).

---

[2] The operator $|\cdot|$ denotes the cardinality of a set.

Proposition 1. *A D2D pair which joins a coalition with the following constraint satisfied will receive a unique payoff in the coalition*

$$\sum_{j \in D_k, j \neq i} \frac{h_{j,i}^k}{h_{i,i}^k \ln 2} < 1, \forall i \in D_k. \tag{9}$$

For detailed proof, see Appendix.

Remark 1. Proposition 1 only shows a sufficient condition of convergence. However, we also note in the simulation that the convergence has a looser constraint than (9). To make the devised algorithm robust, we introduce the maximum iteration number. When the maximum iteration number is reached, the players in the same coalition use their throughput as the payoff. The iterative process in Algorithm 1 can assist D2D pairs to evaluate a coalition.

3.3 Coalition Formation Algorithm for Joint Power and Subchannel Allocation

In the formulated game model, a D2D pair can leave its current coalition and join a new coalition. However, which coalition to choose for the D2D pair remains a challenging problem for the coalition formation game $(\mathcal{N}, V)$. Hence, we define the preference order for the D2D pair to overcome this obstacle.

Definition 2 *(Preference Order). The preference order for a D2D pair i is expressed as $\succ_i$, which is a transitive binary relation over the set of all coalitions a D2D pair i can join.*

The preference order provides a metric to compare which coalition a D2D pair prefers. Consequently, given a D2D pair $i \in \mathcal{N}$ and two coalitions D_k, D_m where $i \in D_k$ and $i \in D_m$, $D_k \succ_i D_m$ means D2D pair i prefers D_k to D_m. Since our aim is to improve the total payoff of D2D pairs, we utilize the utilitarian order (Saad et al. 2009) in this paper.

Definition 3 *(Switch Rule). Given a partition $\pi = \{D_1, D_2, \ldots D_M\}$ of D2D pair set $\mathcal{D}$, a D2D pair i decides to leave its current coalition D_k, $k = 1, 2, \ldots, M$ and join another coalition $D_m \in \pi$, $D_m \neq D_k$, hence forming a new partition π', if only if, $D_m \cup \{i\} \succ_i D_k$, here*

$$D_m \cup \{i\} \succ_i D_k \Leftrightarrow \begin{cases} \sum_{j \in D_k, \forall D_k \in \pi'} v_j \geq \sum_{j \in D_k, \forall D_k \in \pi} v_j \\ v_i\left(D_m \cup \{i\}, \pi'\right) > v_i\left(D_k, \pi\right) \end{cases} \tag{10}$$

where $\pi' = \pi \setminus \{D_k, D_m\} \cup \{D_k \setminus \{i\}, D_m \cup \{i\}\}$; the operator $\Leftrightarrow$ represents left-hand-side and right-hand-side of (10) is equivalent.

The switch rule utilizes utilitarian order. On the right side of (10), the first line implies that payoff of the newly formed partition does not decrease by switching. Meanwhile, the second line indicates the switch operation increases the total payoff of D2D pairs. Since the switch operation of each iteration is only related to coalition D_k and D_m, the following inequations are equivalent

$$\sum_{j\in D_k,\forall D_k\in\pi'} v_j \geq \sum_{j\in D_k,\forall D_k\in\pi} v_j$$

$$\Leftrightarrow \sum_{j\in D_k\cup D_m,D_k,D_m\in\pi'} v_j \geq \sum_{j\in D_k,D_m,D_k,D_m\in\pi} v_j. \tag{11}$$

Furthermore, by applying the switch rule, we present the coalition formation algorithm in Algorithm 2.

Algorithm 2. Coalition Formation With Power Control Algorithm (CFPC)

Initialization
Each D2D pair selects a subchannel randomly and creates the history set $history_i$, $\forall i\in\mathcal{N}$.
Environment discovery
Each D2D pair $i\in\mathcal{N}$ discovers potential coalitions it can join.
Coalition formation process
repeat
    for $i=1:N$ **do**
        D2D pair i lists potential coalitions it is permitted to join, and the current partition is $\pi=\{D_1,D_2,\ldots,D_M\}$.
        D2D pair i negotiates with its potential coalitions, and $v_i(D_k)$ is given in (6), a result of **IPC**.
        D2D pair i decides to join coalition $D_k\in\pi$ based on switch rule in (10) and $D_k\notin history_i$.
    end for
until No D2D pair has incentive to switch
Link level schedule
All D2D pairs in $\mathcal{N}$ start transmit information signal afterwards.

Definition 4. *A partition $\pi=\{D_1,D_2,\ldots,D_M\}$ is called **Nash stable**, if and only if, $\forall i\in\mathcal{N}$, $i\in D_m\in\pi$ such that $D_m\succ_i D_k\cup\{i\}$ for all $D_k\in\pi$.*

Proposition 2. *Starting from any initial network partition π_0, the coalition formation stage of the proposed algorithm always converges to a final Nash Stable parition π^*.*

Proof. Starting from any initial networks partition π_0, there are two possible results after each round of iteration: (1) the network partition is Nash stable; (2) the network partition is not Nash stable. For the first case, the iteration will terminate. For the second case, however, $\exists i\in\mathcal{N}$ with $i\in D_k$ and $D_k,D_m\in\pi$, such that $D_m\cup\{i\}\succ_i D_k$. Therefore, the D2D pair i will conduct switch operation in the next iteration. Since the total number of partition is limited (M^N in our setting) and the proposed algorithm forbids D2D pair revisiting past coalitions, thus all D2D pairs will finally converge to a Nash stable network partition.

4   Simulation Results

In this section, we provide simulation results to illustrate the performances of the proposed CFPC algorithm.

We consider N D2D pairs coexist with M CUEs. Each CUE is assigned an orthogonal subchannel. The transceiver is close enough to satisfy the maximum distance of D2D communication. The channel gain equals to $d^{-\alpha} |h|^2$, where d is the distance between the transceivers, α represents the pathloss factor. The term h denotes the complex Gaussian channel coefficient that satisfies $h \sim \mathcal{CN}(0,1)$. We repeat the simulation 200 times and each time with the newly random-selected locations. We summarize simulation parameters in Table 1.

Table 1. Simulation parameters setting

Parameters	Values
Cell layout	Isolated cell, 1-sector
Cell radius	300 m
Subchannel bandwidth	180 KHz
Noise power	-174 dBm/Hz
Noise Figure	9 dB
TX power	D2D: 23 dBm in maximum, MUE: 23 dBm
Antenna gain	Device: 0 dBi, BS: 14 dBi
The maximum distance of D2D pairs	50 m
Pathloss factor, α	2

Figure 2a shows the sum rate of CUEs and Fig. 2b is the sum rate of D2D pairs. We can see that, as the number of D2D pairs increases, the sum rate of CUEs deceases and the sum rate of D2D pairs increases. When the number of CUE is fixed, more D2D pairs lead to more interference to CUEs, contributing to higher spectrum efficiency for D2D communication. Moreover, Fig. 2a and b illustrate the performance comparison of the proposed algorithm (CFPC) with the one in (Li et al. 2014) with a modification[3] (Classical CF). We can see that the proposed CFPC algorithm outperforms the Classical CF in sum rate of both CUEs and D2D pairs. This is because the proposed scheme enables D2D pairs to further exploit the spectrum reuse gain by power control.

Figure 3 illustrates the fairness performance of the proposed CFPC algorithm and the Classical CF algorithm. We introduce the Jains Fairness index, which is denoted by

---

[3] We allow the D2D pairs to switch their coalition as long as the sum rate of D2D pairs increases in exchange for sum rate of both CUE and D2D pairs rising in (Li et al. 2014).

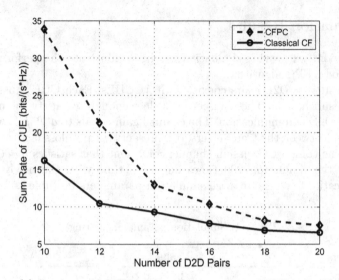

(a) Sum rate of CUEs with varying number of D2D pairs N

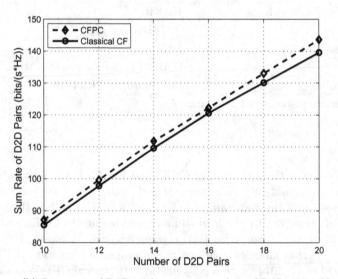

(b) Sum rate of D2D with varying number of D2D pairs N

Fig. 2. Sum rate of CUEs and D2D pairs separately against varying number of D2D pairs.

$$J = \frac{\left(\sum\limits_{i \in \mathcal{M} \cup \mathcal{N}} r_i\right)^2}{(M+N)\sum\limits_{i \in \mathcal{M} \cup \mathcal{N}} r_i^2} \qquad (12)$$

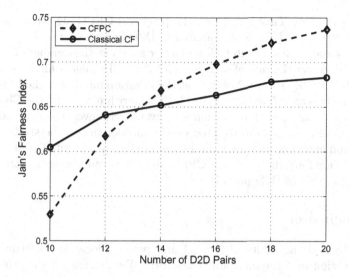

Fig. 3. Jain's Fairness index against the variation of number of D2D pairs

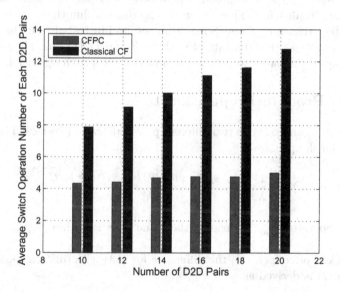

Fig. 4. Average number switch operation of each D2D pair as the number of D2D pairs increases.

as the metric to quantize the fairness. We observe that the Classical CF algorithm offers improved system fairness compared with the proposed scheme when the number of D2D pairs is small. The reason is that power control can lower the transmit power of D2D transmitters, thus can benefit the CUEs significantly compared with that of Classical CF. However, the proposed scheme offers better system fairness compared with the Classical CF algorithm when the number of

D2D pairs is large. That is because the proposed CFPC algorithm can reduce the interference to CUEs as the number of D2D becomes large.

Figure 4 shows that the iteration number grows as the number of D2D pairs increases. However the rate of increasing for the CFPC algorithm is smaller than the Classical CF algorithm. Notice that the iteration number of the Classical CF is significantly larger than that of the CFPC algorithm, because the Classical CF algorithm sacrifices switch operation for fairness. However, the proposed CFPC algorithm had the potential to mitigate the interference in the studied system. This phenomenon becomes more and more obvious as the number of D2D pairs increases; therefore, the proposed CFPC algorithm can obtain a better fairness when the number of D2D pairs is large.

5   Conclusion

We investigated the joint subchannel and power allocation problem for D2D communication underlaying cellular networks. We formulated the problem as a coalition formation game. For the devised game model, a distributed coalition formation algorithm was proposed, where each D2D pair can make decision to leave or join a coalition. We also allowed D2D pairs within the same coalition to optimize their transmit power. Simulation results illustrated that the proposed scheme can increase the sum rate of both CUE and D2D pairs. Meanwhile, the proposed algorithm can also reduce the unnecessary switch operations.

A   The Proof of Proposition 1

Proof. Let $p(1)$ and $p(2)$ be two different power allocation vectors. The solution to (8) can be shown as

$$p_i(m) = \left[\frac{1}{\lambda} - \frac{\sum_{j \in D_k \setminus \{i\}} p_j(m) h_{j,i}^k + N_0 + p_k g_i^k}{h_{i,i}^k} \right]_0^{p_{\max}} \tag{13}$$

where the operator $[x]_0^{p_{\max}}$ denotes the value of x is within $[0, p_{\max}]$, and the $m = 1, 2$.

For a fixed price factor λ, the difference for (13) with different power vector $p(1)$ and $p(2)$ is derived as

$$|p_i(1) - p_i(2)| \leq \left| \sum_{j \in D_k, j \neq i} \frac{h_{j,i}^k}{h_{i,i}^k \ln 2} (p_j(1) - p_j(2)) \right|$$

$$\leq \left(\sum_{j \in D_k, j \neq i} \frac{h_{j,i}^k}{h_{i,i}^k \ln 2} \right) \left| \sum_{j \in D_k, j \neq i} (p_j(1) - p_j(2)) \right| \tag{14}$$

$$< \left| \sum_{j \in D_k, j \neq i} (p_j(1) - p_j(2)) \right|.$$

From (14), we prove that (13) is a non-expansive operator; therefore, we conclude that the iterative procedure will converge to the unique fixed point (Miao et al. 2011, Theorem 3) when $\sum\limits_{j \in D_k, j \neq i} \frac{h_{j,i}^k}{h_{i,i}^k \ln 2} < 1, \forall i \in D_k$.

References

Boccardi, F., Heath, R.W., Lozano, A., Marzetta, T.L., Popovski, P.: Five disruptive technology directions for 5g. IEEE Commun. Mag. **52**(2), 74–80 (2014)

Boyd, S.P., Vandenberghe, L.: Convex Optimization. Cambridge University Press, New York (2004)

Dong, Y., Hossain, M.J., Cheng, J.: Joint power control and time switching for SWIPT systems with heterogeneous qos requirements. IEEE Commun. Lett. **20**(2), 328–331 (2016)

Doppler, K., Rinne, M., Wijting, C., Ribeiro, C.B., Hugl, K.: Device-to-device communication as an underlay to lte-advanced networks. IEEE Commun. Mag. **47**(12), 42–49 (2009)

Feng, D., Lu, L., Wu, Y.-Y., Li, G.Y., Feng, G., Li, S.: Device-to-device communications underlaying cellular networks. IEEE Trans. Commun. **61**(8), 3541–3551 (2013)

Janis, P., Koivunen, V., Ribeiro, C. B., Korhonen, J., Doppler, K., Hugl, K.: Interference-aware resource allocation for device-to-device radio underlaying cellular networks. In: Proceedings of IEEE VTC-Spring, pp. 1–5. Barcelona, April 2009

Lei, L., Zhong, Z., Lin, C., Shen, X.: Operator controlled device-to-device communications in lte-advanced networks. IEEE Wireless Commun. **19**(3), 96–104 (2012)

Li, Y., Jin, D., Yuan, J., Han, Z.: Coalitional games for resource allocation in the device-to-device uplink underlaying cellular networks. IEEE Trans. Wireless Commun. **13**(7), 3965–3977 (2014)

Miao, G., Himayat, N., Li, G.Y., Talwar, S.: Distributed interference-aware energy-efficient power optimization. IEEE Trans. Wireless Commun. **10**(4), 1323–1333 (2011)

Min, H., Lee, J., Park, S., Hong, D.: Capacity enhancement using an interference limited area for device-to-device uplink underlaying cellular networks. IEEE Trans Wireless Commun. **10**(12), 3995–4000 (2011)

Saad, W., Han, Z., Debbah, M., Hjørungnes, A., Başar, T.: Coalitional game theory for communication networks: A tutorial. IEEE Signal Process. Mag. **26**(5), 77–97 (2009)

Tehrani, M.N., Uysal, M., Yanikomeroglu, H.: Device-to-device communication in 5g cellular networks: challenges, solutions, and future directions. IEEE Commun. Mag. **52**(5), 86–92 (2014)

Wang, J., Zhu, D., Zhao, C., Li, J.C.F., Lei, M.: Resource sharing of underlaying device-to-device and uplink cellular communications. IEEE Commun. Lett. **17**(6), 1148–1151 (2013)

Xing, H., Hakola, S.: The investigation of power control schemes for a device-to-device communication integrated into ofdma cellular system. In: Proceedings of IEEE PIMRC, pp. 1775–1780, Instanbul, Sept. 2010

Xu, C., Song, L., Han, Z., Zhao, Q., Wang, X., Cheng, X., Jiao, B.: Efficiency resource allocation for device-to-device underlay communication systems: A reverse iterative combinatorial auction based approach. IEEE J. Sel. Areas Commun. **31**(9), 348–358 (2013)

Xu, C., Song, L., Zhu, D., Lei, M.: Subcarrier and power optimization for device-to-device underlay communication using auction games. In: Proceedings of IEEE ICC, pp. 5526–5531. NSW, Sydney (2014)

Xu, S., Wang, H., Chen, T., Huang, Q., Peng, T.: Effective interference cancellation scheme for device-to-device communication underlaying cellular networks. In: Proceedings of IEEE VTC-Fall, pp. 1–5, Ottawa, ON, Sept. 2010

Yu, C.-H., Doppler, K., Ribeiro, C.B., Tirkkonen, O.: Resource sharing optimization for device-to-device communication underlaying cellular networks. IEEE Trans. Wireless Commun. **10**(8), 2752–2763 (2011)

Yu, C.-H., Tirkkonen, O., Doppler, K., Ribeiro, C.B.: On the performance of device-to-device underlay communication with simple power control. In: Proceedings of IEEE VTC-Spring, pp. 1–5, Barcelona, Apr. 2009a

Yu, C.-H., Tirkkonen, O., Doppler, K., Ribeiro, C.B.: Power optimization of device-to-device communication underlaying cellular communication. In: Proceedings of IEEE ICC, pp. 1–5, Dresden, Jun. 2009b

Author Index

Printed in the United States
By Bookmasters